Lecture Notes in Computer Science 11123

Commenced Publication in 1973
Founding and Former Series Editors:
Gerhard Goos, Juris Hartmanis, and Jan van Leeuwen

More information about this series at http://www.springer.com/series/7407

Igor Potapov · Pierre-Alain Reynier (Eds.)

Reachability Problems

12th International Conference, RP 2018
Marseille, France, September 24–26, 2018
Proceedings

 Springer

Editors
Igor Potapov
University of Liverpool
Liverpool
UK

Pierre-Alain Reynier
Aix-Marseille University
Marseille
France

ISSN 0302-9743 ISSN 1611-3349 (electronic)
Lecture Notes in Computer Science
ISBN 978-3-030-00249-7 ISBN 978-3-030-00250-3 (eBook)
https://doi.org/10.1007/978-3-030-00250-3

Library of Congress Control Number: 2018953208

LNCS Sublibrary: SL1 – Theoretical Computer Science and General Issues

This Springer imprint is published by the registered company Springer Nature Switzerland AG
The registered company address is: Gewerbestrasse 11, 6330 Cham, Switzerland

Preface

This volume contains the papers presented at RP 2018, the 12th International Conference on Reachability Problems, organized on September 24–26, 2018 by Aix-Marseille University, Marseille, France. Previous events in the series were located at: Royal Holloway, University of London (2017), Aalborg University (2016), the University of Warsaw (2015), the University of Oxford (2014), Uppsala University (2013), the University of Bordeaux (2012), the University of Genoa (2011), Masaryk University Brno (2010), École Polytechnique (2009), the University of Liverpool (2008), and Turku University (2007).

The aim of the conference is to bring together scholars from diverse fields with a shared interest in reachability problems, and to promote the exploration of new approaches for the modelling and analysis of computational processes by combining mathematical, algorithmic, and computational techniques. Topics of interest include (but are not limited to): reachability for infinite state systems; rewriting systems; reachability analysis in counter/timed/cellular/communicating automata; Petri nets; computational aspects of semigroups, groups, and rings; reachability in dynamical and hybrid systems; frontiers between decidable and undecidable reachability problems; complexity and decidability aspects; predictability in iterative maps, and new computational paradigms.

Reachability is a fundamental problem that appears in several different contexts. Typically, for a fixed system description given in some form (rewriting rules, transformations by computable functions, systems of equations, logical formulas, etc.) a reachability problem consists in checking whether a given set of target states can be reached starting from a fixed set of initial states. The set of target states can be represented explicitly or via some implicit representation (e.g., a system of equations, a set of minimal elements with respect to some ordering on the states). Sophisticated quantitative and qualitative properties can often be reduced to basic reachability questions. Decidability and complexity boundaries, algorithmic solutions, and efficient heuristics are all important aspects to be considered in this context. Algorithmic solutions are often based on different combinations of exploration strategies, symbolic manipulations of sets of states, decomposition properties, and reduction to linear programming problems, and they often benefit from approximations, abstractions, accelerations, and extrapolation heurisitics. Ad hoc solutions as well as solutions based on general-purpose constraint solvers and deduction engines are often combined in order to balance efficiency and flexibility.

The invited speakers at the RP 2018 were:

- Olivier Bournez - "On the Computational Complexity of Solving Ordinary Differential Equations"
- Maria Prandini - "Reachability in Cyber-Physical Systems"
- Marcin Jurdzinski - "Universal Ordered Trees and Quasi-polynomial Algorithms for Solving Parity Games"

- Jérémie Chalopin - "A Counterexample to Thiagarajan's Conjecture on Regular Event Structures"
- Marta Kwiatkowska - "Safety Verification for Deep Neural Networks with Provable Guarantees"

The conference originally received 29 abstracts from which 21 full papers were submitted. Each submission was carefully reviewed by three Program Committee (PC) members. Based on these reviews, the PC decided to accept 11 papers, in addition to the four invited talks (by Olivier Bournez, Maria Prandini, Marcin Jurdzinski, Jérémie Chalopin) and one invited tutorial (by Marta Kwiatkowska). The members of the PC and the list of external reviewers can be found on the next pages. The PC is grateful for the high quality work produced by these external reviewers. Overall this volume contains 11 contributed papers and the conference also provided the opportunity to other young and established researchers to give informal presentations, prepared shortly before the event, informing the participants about current research and work in progress. The informal presentations have not been included at this LNCS proceedings, but may be found on the conference website.

It is a pleasure to thank the team behind the EasyChair system and the Lecture Notes in Computer Science team at Springer, who together made the production of this volume possible in time for the conference. Finally, we thank all the authors for their high-quality contributions, and the participants for making RP 2018 a success. We are also very grateful to Alfred Hofmann for the continuous support of the event in the last decade and to LNCS Springer, EATCS, CNRS, Laboratoire d'Excellence Archimède, the LIS Laboratory of Computing and Systems, and Aix-Marseille University for the scientific and financial sponsorship of the event.

September 2018

Igor Potapov
Pierre-Alain Reynier

Organization

Program Committee

S. Akshay	IIT Bombay, India
Christel Baier	TU Dresden, Germany
Paul Bell	Liverpool John Moores University, UK
Nathalie Bertrand	Inria, France
Udi Boker	Interdisciplinary Center (IDC) Herzliya, Israel
Krishnendu Chatterjee	Institute of Science and Technology (IST), Austria
Laure Daviaud	The University of Warwick, UK
Giorgio Delzanno	DIBRIS, Università di Genova, Italy
Emmanuel Filiot	Université Libre de Bruxelles, Belgium
Pierre Ganty	IMDEA Software Institute, Spain
Matthew Hague	Royal Holloway University of London, UK
Vesa Halava	University of Turku, Finland
Petr Jancar	Palacky Univ. Olomouc, Czech Republic
Martin Lange	University of Kassel, Germany
Sławomir Lasota	Warsaw University, Poland
Laurent Fribourg	LSV, France
Benjamin Monmege	Aix-Marseille Université, LIF, CNRS, France
Anca Muscholl	LaBRI, Universite Bordeaux, France
Igor Potapov	University of Liverpool, UK
Pavithra Prabhakar	Kansas State University, USA
Alexander Rabinovich	Tel Aviv University, Israel
Pierre-Alain Reynier	Aix-Marseille Université, France
Thomas Schwentick	Universität Dortmund, Germany
Helmut Seidl	Technical University of Munich, Germany
Mikhail Volkov	Ural Federal University, Russia

Additional Reviewers

Balaji, Nikhil
Chillara, Suryajith
Haase, Christoph
Kokkinis, Ioannis

Kwee, Kent
Perez, Guillermo
Totzke, Patrick
Yörük, Lara

Abstracts of Invited Talks

On the Computational Complexity
of Solving Ordinary Differential Equations

Olivier Bournez

Ecole Polytechnique, LIX, 91128 Palaiseau Cedex, France

We consider Continuous Ordinary Differential Equations: That is to say x' = f(x) where $f : \mathbb{R}^n \to \mathbb{R}^n$ is a continuous function. When an initial condition $x(0) = x_0$ is added, this is called an Initial Value Problem (IVP), also called a Cauchy's Problem. A trajectory is any solution of the problem, that is to say, any derivable function $\xi : I \subset \mathbb{R}_{\geq 0} \to \mathbb{R}^n$, where I is some interval containing 0 satisfying $\xi(0) = x_0$, and $\xi'(t) = f(\xi(t))$ on its domain. The solution is said to be maximal, if I is maximal (for inclusion) with this property. For f continuous, IVP are known to always have solutions, but possibly non unique, by Peano-Arzelà's Theorem. When in addition f is Lipschitz (in particular if it is C^1) then unicity is guaranteed, by Cauchy-Lipschitz theorem. When f is analytic, solutions are know to be analytic.

In this talk we will survey various results related to the difficulty of computing a or the solutions for various classes of functions f.

In particular, we will discuss the case $y' = p(t, y)$, $y(t_0) = y_0$, where p is a vector of polynomials). In this case, there is a polynomial time algorithm that, given the initial-value problem, the time T at which we want to compute the solution of the IVP, and the maximum allowable error $\varepsilon > 0$, outputs a value $\tilde{y}_T$ such that $\|\tilde{y}_T - y(T)\| \leq \varepsilon$ in time polynomial in T, $-\log\varepsilon$, and in several quantities related to the polynomial IVP.

We will relate the discussion to questions related to the computational power of several continuous time analog models such as the General Purpose Analog Computer (GPAC) from Claude Shannon. The GPAC was introduced as a model of famous mechanical, and later-on electronics, analog computers named Differential Analysers.

Reachability in Cyber-Physical Systems

Maria Prandini

Politecnico di Milano, Piazza Leonardo da Vinci 32, 20133 Milan, Italy
maria.prandini@polimi.it

Reachability analysis consists in determining the region of the state space that a given dynamical system will visit starting from some set of initial states, subject to a disturbance input modeling uncertainty in the system dynamics and/or the fact that the system is operating in an uncertain environment that can affect its evolution.

A main application of reachability analysis – that makes it relevant to various application domains – is the automatic verification of the correct behavior of a system, which is typically coded by requiring that all its trajectories remain within some desired range of operation and do not enter any forbidden region of the state space. If the outcome of the verification is negative, then, the system has to be redesigned. The availability of some counter-example showing a violation of the correct behavior can be useful to this purpose.

In reachability analysis, the region of the state space that is visited by the system during its evolution is determined by propagating the set of initial states through the uncertain system dynamics, thus computing the so-call reach sets.

The main issue in reachability analysis is indeed the ability to compute with sets. In systems with a finite state space, sets can be represented by enumeration and reach sets can be computed starting from the given initial set and progressively adding one-step successors. If we consider systems involving a continuous state space, then, representation and propagation of reach sets generally become a challenge. One should in fact choose a class of sets that can be efficiently represented and such that, when one applies to these sets the operations involved in their propagation through the system dynamics, then, sets in the same class are obtained. If this is not possible, some outer-approximation of the obtained sets should be adopted to bring their description back to the same class.

Scalability of reach set computations arises as an issue, and calls for abstraction of models through simulation or approximate simulation relations. In the case of a simulation relation, the abstracted model can be used for verifying the correct behavior of the original system since all trajectories of the original system can be generated by simulating the abstracted model (but not vice-versa). For instance, a nonlinear continuous system with smooth dynamics can be reduced to a piecewise affine system that satisfies a simulation relation if the abstraction procedure appropriately accounts for the modeling error through a (fictitious) disturbance input.

We shall consider reachability analysis for cyber-physical systems that represent engineering systems where communication, computation, and control (the cyber part)

Supported by the European Commission under the project UnCoVerCPS with grant number 643921.

are integrated within natural and/or human-made systems (the physical part) governed by the laws of physics. Hybrid models are used to describe this class of systems, since the interleaved discrete and continuous state components of a hybrid model can represent the cyber and physical parts integrated in a cyber-physical system.

Reachability analysis of hybrid systems is challenging since their hybrid state has a continuous component and the propagation of the reach sets in the continuous state space depends on the value taken by the hybrid state. Typically, a reach set in the continuous state space can split in subsets that propagate according to different continuous dynamics, thus growing the effort in reach set computations.

In this invited talk, we shall focus on discrete time piecewise affine systems, which often arise as a model for cyber-physical systems and have also some potential as a unifying modeling framework for automatic verification of nonlinear continuous systems. More specifically, we address verification of discrete time piecewise affine systems based on reach set computations, including the generation of counter-examples, and the use of abstraction and invariant sets to improve scalability. We also address the case when a control input is available to impose the correct system behavior via disturbance compensation, and describe a set-based approach to feedback control design integrating reach set computations.

Universal Trees and Quasi-Polynomial Algorithms for Solving Parity Games

Marcin Jurdziński

Department of Computer Science, University of Warwick, UK

Parity games have played a fundamental role in automata theory, logic, and their applications to verification and synthesis since early 1990's. Solving parity games is polynomial-time equivalent to checking *emptiness of automata on infinite trees* and to the *modal mu-calculus model checking*. It is a long-standing open question whether there is a polynomial-time algorithm for solving parity games. The quest for a polynomial-time algorithm has not only brought diverse algorithmic techniques to the theory and practice of verification and synthesis, but it has also significantly contributed to resolving long-standing open problems in other research areas, such as Markov Decision Processes and Linear Programming.

All algorithms for solving parity games that were known until 2016 required time that was exponential in the most important parameter of a parity game—the number of distinct *priorities*. The major breakthrough was achieved by Calude, Jain, Khoussainov, Li, and Stephan in 2017, who have given the first quasi-polynomial algorithm and established that parity games are in FPT (fixed-parameter tractable). Two other quasi-polynomial algorithms for solving parity games were subsequently devised by Jurdziński and Lazić, 2017, and by Lehtinen, 2018, and a space-efficient version of Calude et al.'s algorithm was given by Fearnley, Jain, Schewe, Stephan, and Wojtczak, 2017. The conceptual and technical toolkits used by all the three algorithms seem rather distinct: the breakthrough result of Calude et al. was based on computing *play summaries* by *succinct counting*, Jurdziński and Lazić have devised a *succinct coding* of *ordered trees* and applied it to the *progress measure lifting* algorithm, and Lehtinen has developed novel concepts of *register games* and the *register index*.

In this talk we first focus on presenting the technical insights of the quasi-polynomial algorithm for solving parity games that is based on progress measure lifting and succinct coding of ordered trees. Following Czerwiński, Daviaud, Fijalkow, Jurdziński, Lazić, and Parys, 2018, we then argue that *universal ordered trees*— implicit in the succinct tree-coding result of Jurdziński and Lazić—offer a unifying perspective on the three distinct quasi-polynomial algorithms. Moreover, the analysis of universal trees leads to an automata-theoretic quasi-polynomial lower bound that forms a barrier that all the existing approaches, as well as other possible techniques that follow the separation approach, must overcome in the quest for a polynomial-time algorithm for solving parity games.

More specifically, we argue that the techniques underlying all the three quasi-polynomial algorithms can be interpreted as constructions of automata on infinite words that are of quasi-polynomial size and that facilitate solving parity games by the *separation approach* formalized by Bojańczyk and Czerwiński, 2018, and implicit in the work of Bernet, Janin, and Walukiewicz, 2002. In particular, we point out how such *separating automata* arise in a very natural way from universal ordered trees. Then we present two lower bounds: one is a quasi-polynomial lower bound on the size of universal trees that nearly matches (up to a small polynomial factor) the succinct tree-coding upper bound of Jurdziński and Lazić, and the other establishes that the set of states in every separating automaton contains leaves of some universal tree, which implies that every separating automaton is of at least quasi-polynomial size.

Keywords: Parity games · Quasi-polynomial algorithms · Progress measures Universal ordered trees · Separating automata · Lower bounds

A Counterexample to Thiagarajan's Conjecture on Regular Event Structures

Jérémie Chalopin

LIS, CNRS, Aix-Marseille Université, and Universit de Toulon
jeremie.chalopin@lis-lab.fr

We provide a counterexample to a conjecture by Thiagarajan [8, 9] that regular event structures correspond exactly to event structures obtained as unfoldings of finite 1-safe Petri nets. Event structures, trace automata, and Petri nets are fundamental models in concurrency theory. There exist nice interpretations of these structures as combinatorial and geometric objects and both conjectures can be reformulated in this framework. Namely, from a graph theoretical point of view, the domains of prime event structures correspond exactly to median graphs; from a geometric point of view, these domains are in bijection with CAT(0) cube complexes.

A necessary condition for the conjecture to be true is that domains of regular event structures admit a regular nice labeling (which corresponds to a special coloring of the hyperplanes of the associated CAT(0) cube complex). To disprove these conjectures, we describe a regular event domain that does not admit a regular nice labeling. Our counterexample is derived from an example by Wise [10, 11] of a nonpositively curved square complex $\mathbf{X}$ with six squares, whose edges are colored in five colors, and whose universal cover $\widetilde{\mathbf{X}}$ is a CAT(0) square complex containing a particular plane with an aperiodic tiling. We prove that other counterexamples to Thiagarajan's conjecture arise from aperiodic 4-way deterministic tile sets of Kari and Papasoglu [6] and Lukkarila [7].

On the positive side, we show that event structures obtained as unfoldings of finite 1-safe Petri nets correspond to the finite special cube complexes. This subclass of nonpositively curved cube complexes was introduced by Haglund and Wise [4, 5] in geometric group theory and is characterized by simple combinatorial properties satisfied by the hyperplanes. Using the breakthrough results by Agol [1] based on special cube complexes, we prove that Thiagarajan's conjecture is true for regular event structures whose domains occur as principal filters of hyperbolic CAT(0) cube complexes which are universal covers of finite nonpositively curved cube complexes.

Joint work with Victor Chepoi.

The full version of this paper is available on ArXiv [2], an extended abstract appeared in the proceedings of ICALP 2017 [3].

References

1. Agol, I.: The virtual Haken conjecture. Doc. Math. **18**, 1045–1087 (2013). with an appendix by Ian Agol, Daniel Groves, and Jason Manning
2. Chalopin, J., Chepoi, V.: A counterexample to Thiagarajan's conjecture on regular event structures. arXiv preprint (2016)
3. Chalopin, J., Chepoi, V.: A counterexample to Thiagarajan's conjecture on regular event structures. In: ICALP. LIPIcs, vol. 80, pp. 101:1–101:14. Schloss Dagstuhl -Leibniz-Zentrum für Informatik (2017)
4. Haglund, F., Wise, D.: Special cube complexes. Geom. Funct. Anal. **17**(5), 1551–1620 (2008)
5. Haglund, F., Wise, D.: A combination theorem for special cube complexes. Annals Math. **176** (3), 1427–1482 (2012)
6. Kari, J., Papasoglu, P.: Deterministic aperiodic tile sets. GAFA, Geom. Funct. Anal. **9**(2), 353–369 (1999)
7. Lukkarila, V.: The 4-way deterministic tiling problem is undecidable. Theor. Comput. Sci. **410**(16), 1516–1533 (2009)
8. Thiagarajan, P.: Regular trace event structures. Technical report BRICS RS-96-32, Computer Science Department, Aarhus University, Aarhus, Denmark (1996)
9. Thiagarajan, P.: Regular event structures and finite petri nets: a conjecture. In: Brauer, W., Ehrig, H., Karhumäki, J., Salomaa, A. (eds.) Formal and Natural Computing. LNCS, vol. 2300, pp. 244–256. Springer, Heidelberg (2002)
10. Wise, D.: Non-positively curved squared complexes, aperiodic tilings, and non-residually finite groups. Ph.D. thesis, Princeton University (1996)
11. Wise, D.: Complete square complexes. Comment. Math. Helv **82**(4), 683–724 (2007)

Safety Verification for Deep Neural Networks with Provable Guarantees (Extended Abstract)

Marta Kwiatkowska

Department of Computing Science, University of Oxford, UK

Deep neural networks have achieved impressive experimental results in image classification, but can surprisingly be unstable with respect to adversarial perturbations, that is, minimal changes to the input image that cause the network to misclassify it. With potential applications including perception modules and end-to-end controllers for self-driving cars, this raises concerns about their safety. This lecture will describe progress with developing automated verification techniques for deep neural networks to ensure safety of their classification decisions with respect to image manipulations, for example scratches or changes to camera angle or lighting conditions, that should not affect the classification. The techniques exploit Lipschitz continuity of the networks and aim to approximate, for a given set of inputs, the reachable set of network outputs in terms of lower and upper bounds, in anytime manner, with provable guarantees. We develop novel algorithms based on games and global optimisation, and evaluate them on state-of-the-art networks.

Robustness of neural networks is an active topic of investigation and a number of approaches have been proposed to search for adversarial examples. They are based on computing the gradients [1, 3], computing a Jacobian-based saliency map [6], transforming the existence of adversarial examples into an optimisation problem [2], and transforming the existence of adversarial examples into a constraint solving problem [5]. In contrast, this lecture reports on research that aims to rule out the existence of adversarial examples, which approaches based on heuristic search are not able to achieve. In particular, we will adopt the definition of safety based on pointwise robustness introduced in [4], where the first practical automated verification method was developed, based on discretising the neighbourhood and searching it exhaustively in a layer-by-layer manner. A brief overview will also be given of two approaches that utilise Lipschitz continuity, one based on global optimisation [7], and capable of expressing the safety of [4] as well as reachability, and the other [8, 9] on reducing dimensionality by working with black or grey box feature extraction and searching for adversarial examples using a two-player game, where the first player targets the features and the second targets pixels within the feature. The game tree is traversed using Monte Carlo tree search and variants of A* and Alpha-Beta pruning, which produces successive lower and upper bounds on the maximum safe radius with asymptotic convergence guarantees.

References

1. Biggio, B., et al.: Evasion attacks against machine learning at test time. In: Blockeel, H., Kersting, K., Nijssen, S., Železný, F. (eds.) ECML PKDD 2013. LNCS, vol. 8190, pp. 387–402. Springer, Heidelberg (2013)
2. Nicholas, C., David, W.: Towards evaluating the robustness of neural networks. In: 2017 IEEE Symposium on Security and Privacy (SP), pp. 39–57. IEEE (2017)
3. Goodfellow, I.J., Shlens, J., Szegedy, C.: Explaining and harnessing adversarial examples. CoRR
4. Huang, X., Kwiatkowska, M., Wang, S., Wu, M.: Safety verification of deep neural networks. In: Majumdar, R., Kunčak, V. (eds.) CAV 2017. LNCS, vol. 10426, pp. 3–29. Springer, Cham (2017)
5. Katz, G., Barrett, C., Dill, D.L., Julian, K., Kochenderfer, M.J.: Reluplex: an efficient SMT solver for verifying deep neural networks. In: Majumdar, R., Kunčak, V. (eds.) CAV 2017. LNCS, vol. 10426, pp. 97–117. Springer, Cham (2017)
6. Papernot, N., McDaniel, P., Jha, S., Fredrikson, M., Celik, Z.B., Swami, A: The limitations of deep learning in adversarial settings. In: 2016 IEEE European Symposium on Security and Privacy (EuroS&P), pp. 372–387. IEEE (2016)
7. Ruan, W., Huang, X., Kwiatkowska, M.: Reachability analysis of deep neural networks with provable guarantees. In: International Joint Conference on Artificial Intelligence (2018)
8. Wicker, M., Huang, X., Kwiatkowska, M.: Feature-guided black-box safety testing of deep neural networks. In: Beyer, D., Huisman, M. (eds.) TACAS 2018. LNCS, vol. 10805, pp. 408–426. Springer, Cham (2018)
9. Wu, M., Wicker, M., Ruan, W., Huang, X., Kwiatkowska, M.: A game-based approximate verification of deep neural networks with provable guarantees. CoRR, abs/1807.03571 (2018)

Contents

Reachability Analysis of Nonlinear ODEs Using Polytopic Based Validated
Runge-Kutta.. 1
   Julien Alexandre dit Sandretto and Jian Wan

The Satisfiability of Word Equations: Decidable and Undecidable
Theories... 15
   Joel D. Day, Vijay Ganesh, Paul He, Florin Manea,
   and Dirk Nowotka

Left-Eigenvectors Are Certificates of the Orbit Problem............... 30
   Steven de Oliveira, Virgile Prevosto, Peter Habermehl,
   and Saddek Bensalem

Constrained Dynamic Tree Networks............................. 45
   Matthew Hague and Vincent Penelle

EXPSPACE-Complete Variant of Countdown Games, and Simulation
on Succinct One-Counter Nets.................................. 59
   Petr Jančar, Petr Osička, and Zdeněk Sawa

Revisiting MU-Puzzle. A Case Study in Finite Countermodels
Verification... 75
   Alexei Lisitsa

Knapsack in Hyperbolic Groups................................ 87
   Markus Lohrey

Generalized Tag Systems...................................... 103
   Turlough Neary and Matthew Cook

Certain Query Answering on Compressed String Patterns: From Streams
to Hyperstreams.. 117
   Iovka Boneva, Joachim Niehren, and Momar Sakho

Büchi VASS Recognise Σ_1^1-complete ω-languages................... 133
   Michał Skrzypczak

Qualitative Reachability for Open Interval Markov Chains.............. 146
   Jeremy Sproston

Author Index 161

Reachability Analysis of Nonlinear ODEs Using Polytopic Based Validated Runge-Kutta

Julien Alexandre dit Sandretto[1(✉)] and Jian Wan[2]

[1] U2IS, ENSTA ParisTech, 828 bd des Maréchaux, 91762 Palaiseau, France
`alexandre@ensta.fr`
[2] School of Engineering, University of Plymouth, Plymouth, Devon PL4 8AA, UK
`jian.wan@plymouth.ac.uk`

Abstract. Ordinary Differential Equations (ODEs) are a general form of differential equations. This mathematical format is often used to represent the dynamic behavior of physical systems such as control systems and chemical processes. Linear ODEs can usually be solved analytically while nonlinear ODEs may need numerical methods to obtain approximate solutions. There are also various developments for validated simulation of nonlinear ODEs such as explicit and implicit guaranteed Runge-Kutta integration schemes. The implicit ones are mainly based on zonotopic computations using affine arithmetics. It allows to compute the reachability of a nonlinear ODE with a zonotopic set as its initial value. In this paper, we propose a new validated approach to solve nonlinear ODEs with a polytopic set as the initial value using an indirectly implemented polytopic set computation technique.

1 Introduction

Many scientific applications such as those in mechanics, robotics, chemistry and electronics require the solution of ordinary differential equations (ODEs). In the general case, nonlinear ODEs can not be solved analytically and a numerical integration scheme is used instead to obtain approximate solutions. Nevertheless, for some applications as in [3,6,9,15], an approximation of the solution is not sufficient and a bound for the exact solution is mandatorily required.

The problem to be studied here is about the computation of the solution for the *set initial value problem (SIVP)* of an autonomous *Ordinary Differential Equation* defined as follows:

$$\dot{\mathbf{y}} = \mathbf{f}(\mathbf{y}) \quad \text{with} \quad \mathbf{y}(0) \in \mathcal{Y}_0 \quad \text{and} \quad t \in [0, t_{\text{end}}]. \tag{1}$$

The function $\mathbf{f} : \mathbb{R}^n \to \mathbb{R}^n$ is assumed to be nonlinear, $\mathbf{y} \in \mathbb{R}^n$ is the vector of state variables, and $\dot{\mathbf{y}}$ is the derivative of $\mathbf{y}$ with respect to time t. The function $\mathbf{f}$ is also assumed to be globally Lipschitz in $\mathbf{y}$ for Eq. (1) to have a unique solution under the initial condition $\mathbf{y}_0$ [8]. Furthermore, the function $\mathbf{f}$ is further assumed

I. Potapov and P.-A. Reynier (Eds.): RP 2018, LNCS 11123, pp. 1–14, 2018.
https://doi.org/10.1007/978-3-030-00250-3_1

to be continuously differentiable. Note that the initial value is given as a set, *i.e.*, there are some bounded uncertainties for the initial value. So the solution for the corresponding problem is $\mathbf{y}(t; \mathcal{Y}_0)$, which is defined as follows:

$$\mathbf{y}(t; \mathcal{Y}_0) = \{\mathbf{y}(t; \mathbf{y}_0) : \mathbf{y}_0 \in \mathcal{Y}_0\}.$$

The solution of Eq. (1) cannot be computed straightforwardly for a general set $\mathcal{Y}_0$. An alternative approach is to bound the set of initial values in a box $\mathcal{Y}_0 \subset [\mathbf{y}_0]$ and then to use interval arithmetic to solve the resulting problem:

$$\dot{\mathbf{y}} = \mathbf{f}(\mathbf{y}) \quad \text{with} \quad \mathbf{y}(0) \in [\mathbf{y}_0] \quad \text{and} \quad t \in [0, t_{\text{end}}]. \tag{2}$$

There exist several approaches to solve the above problem such as those in [1,2,11–13]. However, the initial set for these approaches is constrained to be a box and thus these approaches are rather limited in terms of flexibility and accuracy. In fact, the initial set can also be represented by a zonotopic set, which has a more flexible shape than a box. Accordingly, zonotopic set computation has been used instead to solve nonlinear ODEs with a zonotopic set as the initial set [1,2]. Zonotopic set computation can be implemented using affine arithmetic [5].

Since polytopes are the most common convex sets resulting from linear inequalities, it is more often to encounter an ODE with a polytopic set as the initial set. Unlike zonotopes or boxes, the propagation of a polytopic set for a nonlinear system cannot be computed directly. Using an indirectly implemented polytopic set computation technique proposed in [16], the above problem can be extended naturally to the case of having a polytopic set as the initial set.

The paper is organized as follows. In Sect. 2, the existing validated Runge-Kutta method with a zonotopic set as the initial value is introduced. In Sect. 3, the indirectly implemented polytopic set computation technique is described to extend the existing validated Runge-Kutta method in terms of the shape for the initial set. Section 4 proposes the extended validated Runge-Kutta method with a polytopic set as the initial value and two illustrative examples are provided to demonstrate the main contribution of the paper. Finally, some conclusions are given in Sect. 5.

Notations. x denotes a real value while $\mathbf{x}$ represents a vector of real values. $[x]$ represents an interval value. An interval $[x_i] = [\underline{x_i}, \overline{x_i}]$ defines the set of reals x_i such that $\underline{x_i} \le x_i \le \overline{x_i}$. $\mathbb{IR}$ denotes the set of all intervals while $\mathbb{R}$ denotes the set of real values. The size or the width of $[x_i]$ is $w([x_i]) = \overline{x_i} - \underline{x_i}$ and $\text{m}([x])$ denotes the center of $[x]$. A vector of intervals, or a *box*, $[\mathbf{x}]$ is the Cartesian product of intervals $[x_1] \times \ldots \times [x_i] \times \ldots \times [x_n]$.

2   Zonotopic Based Validated Runge-Kutta

2.1   Initial Value Problem

Runge-Kutta methods can solve the *initial value problem* (*IVP*) of non-autonomous ODEs defined by

$$\dot{\mathbf{y}} = \mathbf{f}(t, \mathbf{y}) \quad \text{with} \quad \mathbf{y}(0) = \mathbf{y}_0 \quad \text{and} \quad t \in [0, t_{\text{end}}]. \tag{3}$$

The function $\mathbf{f} : \mathbb{R} \times \mathbb{R}^n \rightarrow \mathbb{R}^n$ is called the *vector field*, $\mathbf{y} \in \mathbb{R}^n$ is called the *vector of state variables*, and $\dot{\mathbf{y}}$ denotes the derivative of $\mathbf{y}$ with respect to time t. $\mathbf{f}$ is assumed to be globally Lipschitz in $\mathbf{y}$, so Eq. (3) admits a unique solution for a given initial condition $\mathbf{y}_0$ [8]. $\mathbf{f}$ is also assumed to be continuously differentiable. The exact solution of Eq. (3) is denoted by $\mathbf{y}(t; \mathbf{y}_0)$, often called the *flow*.

2.2 Validated Runge-Kutta

The goal of the numerical solution to Eq. (3) is to compute a sequence of time instants $0 = t_0 < t_1 < \cdots < t_N = t_{\text{end}}$ and a sequence of states $\mathbf{y}_0, \ldots, \mathbf{y}_N$ such that $\forall \ell \in [0, N]$, $\mathbf{y}_\ell \approx \mathbf{y}(t_\ell, \mathbf{y}_{\ell-1})$, which is to be obtained by an integration scheme.

A Runge-Kutta method, starting from an initial value $\mathbf{y}_\ell$ at time t_ℓ and a finite time horizon h, the *step size*, produces an approximation $\mathbf{y}_{\ell+1}$ at time $t_{\ell+1}$, with $t_{\ell+1} - t_\ell = h$, of the solution $\mathbf{y}(t_{\ell+1}; \mathbf{y}_\ell)$. Furthermore, to compute $\mathbf{y}_{\ell+1}$, a Runge-Kutta method computes s evaluations of f at predetermined time instants. The number s is known as the number of *stages* of a Runge-Kutta method. More precisely, a Runge-Kutta method is defined by

$$\mathbf{y}_{\ell+1} = \mathbf{y}_\ell + h \sum_{i=1}^{s} b_i \mathbf{k}_i, \tag{4}$$

with $\mathbf{k}_i$ defined by

$$\mathbf{k}_i = \mathbf{f}\left(t_\ell + c_i h, \mathbf{y}_\ell + h \sum_{j=1}^{s} a_{ij} \mathbf{k}_j \right). \tag{5}$$

The coefficients c_i, a_{ij} and b_i, for $i, j = 1, 2, \cdots, s$, fully characterize the Runge-Kutta methods, and they are usually synthesized in a *Butcher tableau* [4] of the form

$$
\begin{array}{c|cccc}
c_1 & a_{11} & a_{12} & \ldots & a_{1s} \\
c_2 & a_{21} & a_{22} & \ldots & a_{2s} \\
\vdots & \vdots & \vdots & \ddots & \vdots \\
c_s & a_{s1} & a_{s2} & \ldots & a_{ss} \\
\hline
 & b_1 & b_2 & \ldots & b_s
\end{array}
\quad \equiv \quad
\begin{array}{c|c}
\mathbf{c} & \mathbf{A} \\
\hline
 & \mathbf{b}
\end{array}.
$$

For example, the Butcher tableau of the well-known RK4 method is given by

$$
\begin{array}{c|cccc}
0 & 0 & 0 & 0 & 0 \\
\frac{1}{2} & \frac{1}{2} & 0 & 0 & 0 \\
\frac{1}{2} & 0 & \frac{1}{2} & 0 & 0 \\
1 & 0 & 0 & 1 & 0 \\
\hline
 & \frac{1}{6} & \frac{1}{3} & \frac{1}{3} & \frac{1}{6}
\end{array}
\tag{6}
$$

To make Runge-Kutta validated [1], the challenging question is how to compute a bound on the difference between the true solution and the numerical solution, defined by $\mathbf{y}(t_\ell; \mathbf{y}_{\ell-1}) - \mathbf{y}_\ell$. This distance is associated to the *local truncation error* (LTE) of the numerical method. It has been shown that LTE can be easily bounded by using the difference between the Taylor series of the exact and the numerical solutions, which is reduced to be LTE $= \mathbf{y}^{(p+1)}(t_\ell) - [\mathbf{y}_\ell^{(p+1)}]$, that is to say the difference of the $(p+1)^{th}$ Taylor coefficients, with p the order of the considered method. This difference has to be evaluated on a specific box, obtained with the Picard-Lindelöf operator, but this is out of the scope of this paper, see [1] for more details. For a method with interval coefficients, the LTE is bounded with guarantee (even over-approximated), which is not the case for a method with floating-point coefficients. For a validated method, the use of interval coefficients is therefore a requirement.

The problem of IVPs with the initial value given in a set is to be considered in this paper. Validated Runge-Kutta approach works well with Interval IVPs (IIVP), *i.e.*with $\mathbf{y}_0 \in [\mathbf{y}_0]$.

2.3   Affine Arithmetic

In order to avoid or to limit the conservativeness from the dependency problem of intervals, *affine arithmetic* [5,14] is to be used instead of interval arithmetic for validated Runga-Kutta. Affine arithmetic can track linear correlations among state variables. A set of values in this domain is represented by an *affine form* $\hat{x}$, which is a formal expression of the form $\hat{x} = \alpha_0 + \sum_{i=1}^{n} \alpha_i \varepsilon_i$ where the coefficients α_i are real numbers, α_0 being called the *center* of the affine form, and the ε_i are formal variables ranging over the interval $[-1, 1]$. Obviously, an interval $a = [a_1, a_2]$ can be represented by the affine form $\hat{x} = \alpha_0 + \alpha_1 \varepsilon$ with $\alpha_0 = (a_1 + a_2)/2$ and $\alpha_1 = (a_2 - a_1)/2$. Moreover, affine forms encode linear dependencies among variables: if $x \in [a_1, a_2]$ and y is such that $y = 2x$, then x will be represented by the affine form $\hat{x}$ above and y will be represented as $\hat{y} = 2\alpha_0 + 2\alpha_1 \varepsilon$.

Usual operations on real numbers extend to affine arithmetic in the expected way. For instance, if $\hat{x} = \alpha_0 + \sum_{i=1}^{n} \alpha_i \varepsilon_i$ and $\hat{y} = \beta_0 + \sum_{i=1}^{n} \beta_i \varepsilon_i$, then with $a, b, c \in \mathbb{R}$ we have

$$a\hat{x} + b\hat{y} + c = (a\alpha_0 + b\beta_0 + c) + \sum_{i=1}^{n}(a\alpha_i + b\beta_i)\varepsilon_i.$$

However, unlike the addition, most operations create new noise symbols. Multiplication for example is defined by

$$\hat{x} \times \hat{y} = \alpha_0 \alpha_1 + \sum_{i=1}^{n}(\alpha_i \beta_0 + \alpha_0 \beta_i)\varepsilon_i + \nu \varepsilon_{n+1},$$

where $\nu = (\sum_{i=1}^{n} |\alpha_i|) \times (\sum_{i=1}^{n} |\beta_i|)$ over-approximates the error between the linear approximation of multiplication and multiplication itself. Example 2.1 illustrates the benefit of affine arithmetic.

Example 2.1. Consider again $e = x + h \times (-x)$ with $h = 0.5$ and $x = [0,1]$ which is associated to the affine form $\hat{x} = 0.5 + 0.5\varepsilon_1$. Evaluating e with affine arithmetic without rewriting the expression, we obtain $[0,0.5]$ as a result.   ■

Example 2.1 also shows the important role of affine arithmetic when it is combined with numerical integration methods. Most of all, it shows the necessity to keep track of the linear dependency between state variables in order to reduce the conservativeness.

Other operations, like sin, exp, are evaluated using either the Min-Range method or a Chebychev approximation, see [5,14] for more details.

2.4   Zonotopes

Considering m affine forms $\hat{x}^1, \ldots, \hat{x}^m$, a joint range $\langle \hat{x}^1, \ldots, \hat{x}^m \rangle \subset \mathbb{R}^m$ is defined as the set of all tuples $(x^1, \ldots, x^m)$ of values compatible with those affine forms. The set defined by $\langle \hat{x}^1, \ldots, \hat{x}^n \rangle$ is the parallel projection on $\mathbb{R}^m$ of the hypercube $\mathbb{U}^n$ by the affine map $(\hat{x}^1, \ldots, \hat{x}^m)$. The projection is a *zonotope*, a center-symmetric convex polytope in $\mathbb{R}^m$.

2.5   Scheme with Affine Arithmetic

With this affine arithmetic, the initial value can be taken in a zonotope Z_0, such that $y_0 \in Z_0 = \alpha_0 + \sum_{i=1}^{n} \alpha_i \varepsilon_i$.

Then, after time elapsed h with RK4 scheme:

$$
\begin{bmatrix}
\mathbf{k}_1 = \mathbf{f}(\mathbf{y}_0) \\
\mathbf{k}_2 = \mathbf{f}(\mathbf{y}_0 + 0.5h\mathbf{k}_1) \\
\mathbf{k}_3 = \mathbf{f}(\mathbf{y}_0 + 0.5h\mathbf{k}_2) \\
\mathbf{k}_4 = \mathbf{f}(\mathbf{y}_0 + h\mathbf{k}_3) \\
\mathbf{y}(h) = \mathbf{y}_0 + h(1/6\mathbf{k}_1 + 1/3\mathbf{k}_2 + 1/3\mathbf{k}_3 + 1/6\mathbf{k}_4) + \text{LTE}
\end{bmatrix} \tag{7}
$$

These terms have to be evaluated with affine arithmetic, except the LTE which is computed with interval arithmetic.

Running Example. Consider the Volterra system given by the following equation:

$$
\begin{cases}
\dot{y}_1 = 2y_1(1 - y_2) \\
\dot{y}_2 = -y_2(1 - y_1)
\end{cases} \tag{8}
$$

with initial conditions: $Z_0 = \begin{bmatrix} 1.0 \\ 3.0 \end{bmatrix} \oplus \begin{bmatrix} 0.04 \, 0.02 \\ 0.02 \, 0.02 \end{bmatrix} B^2$.

We integrate one step with $h = 0.00245$, the computed LTE is then equal to the box $([-7.6e-13, -4.6e-13]; [3.1e-13, 4.6e-13])$ (which is under the chosen tolerance of $1e-10$). Then, with the evaluation of Eq. (7) provides the solution:

$$
\mathbf{y}(h) \in \begin{bmatrix} 0.9902440 \\ 2.9999640 \end{bmatrix} \oplus \begin{bmatrix} 0.0395119 \; 0.0197074 \\ 0.0202920 \; 0.02014574 \end{bmatrix} B^2 + \left(\begin{bmatrix} [-2e-06, 2e06] \\ [-2e-06, 2e06] \end{bmatrix} \right)
$$

The box added to the zonotope gathers the small noises, compacted to reduce the number of generators, coming from the nonlinear operations.

2.6   If Integration Fails

In validated numerical integration, it can happen that the integration fails. Often, it is due to the too large LTE compared to the chosen tolerance, or because the Picard operator itself fails (the existence and uniqueness cannot be proved, see [1] for details).

In this case, it is common to reduce the step size h. Another method is the bisection of the initial conditions such that $Z_0 = Z_0^1 \cup Z_0^2$. After that, two simulations are launched to obtain two solutions such that $\mathbf{y}(Z_0, t_f) = \mathbf{y}(Z_0^1, t_f) \cup \mathbf{y}(Z_0^2, t_f)$.

This is a recursive approach: if simulation from Z_0^2 fails as well, then $\mathbf{y}(Z_0, t_f) = \mathbf{y}(Z_0^1, t_f) \cup (\mathbf{y}(Z_0^{21}, t_f) \cup \mathbf{y}(Z_0^{22}, t_f))$.

Running Example. We reuse the example given in Sect. 2.5. Validated simulation cannot reach the required solution at $t_f = 6$, and fails at $t = 5.783$. The obtained solution is then $\mathbf{y}(Z_0, t_f) = ([-\infty, \infty]; [-\infty, \infty])$ (which is correct anyway). Using the bisection method presented in Sect. 3, we consider the two zonotopes described by $Z_0^1 = \begin{bmatrix} 0.98 \\ 2.99 \end{bmatrix} \oplus \begin{bmatrix} 0.02 \ 0.02 \\ 0.01 \ 0.02 \end{bmatrix} B^2$, and $Z_0^2 = \begin{bmatrix} 1.02 \\ 3.01 \end{bmatrix} \oplus \begin{bmatrix} 0.02 \ 0.02 \\ 0.01 \ 0.02 \end{bmatrix} B^2$. The initial conditions after bisections are given in Fig. 1, and the results of reachability, after the union, are given in Fig. 2.

3   Polytope Geometry

Polytope is a bounded polyhedron $\mathcal{P} \subset \mathbb{R}^n$, which can be defined as follows:

$$\mathcal{P} = \{ \mathbf{x} \in \mathbb{R}^n | H\mathbf{x} \leqslant \mathbf{k} \}, \tag{9}$$

where H is a matrix of $m \times n$ and $\mathbf{k}$ is a column vector of dimension m. Basic polytope manipulations such as the intersection of polytopes and the convex hull for a union of polytopes are implemented in Multi-Parametric Toolbox [10].

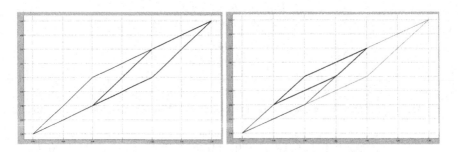

Fig. 1. Initial conditions of running example after one bisection (left) and two bisections (right).

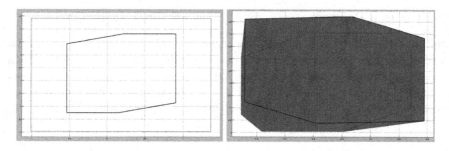

Fig. 2. Solution at $t = 6$ of running example after one bisection (left: two zonotopes, one is fully included in the other one, a box) and two bisections (right: two zonotopes, the union is the polytope in red). (Color figure online)

It is worthy to note that a zonotope is a centrally symmetric polytope. The explicit representation of a zonotope or the representation of a zonotope in the format of a polytope is the zonotope construction problem, which aims to list all extreme points of a zonotope defined by its line segment generators. The zonotope construction problem was addressed in [7], where the addition of line segments was replaced by the addition of polytopes.

3.1 Represent a Polytope Exactly by the Intersection of Zonotopes

As shown in (7), the dynamic evolution of a nonlinear system with a zonotope as the initial state can be computed directly using affine arithmetic. If the initial set is a polytope, there is no direct method to compute the dynamic evolution of this nonlinear system with the polytope as the initial state since its mathematical format involves inequality constraints. However, a polytope can be represented exactly by the intersection of zonotopes as proposed in [16]. Once the polytope $\mathcal{P}$ has been represented exactly by the intersection of zonotopes, i.e., $\mathcal{P} = Z_1 \cap \cdots \cap Z_n$, the dynamic evolution of the nonlinear system with the polytope as the initial state can be computed as follows:

$$f(\mathcal{P}) = f(Z_1 \cap \cdots \cap Z_n) \subseteq f(Z_1) \cap \cdots \cap f(Z_n), \tag{10}$$

where $f(Z_1), \cdots, f(Z_n)$ can be computed using affine arithmetic and the intersection of zonotopes can be transformed to be the intersection of the constructed polytopes.

As explained in [16], the general procedures to represent a polytope $\mathcal{P} \subset \mathbb{R}^n$ exactly by the intersection of zonotopes are: randomly select n inequality constraints from the pool of all inequality constraints for the polytope and then to use these n inequality constraints to construct a zonotope with the minimized volume to contain the polytope until all inequality constraints for the polytope have been used up.

3.2   Bisect a Polytope

Similar to the bisection of a zonotope in case of failed integration, the initial set of a polytope can also be bisected into two sub-polytopes. Any line passing through the Chebyshev center of the polytope can be used to bisect the polytope. Taking the following 2-D polytope in Fig. 3 as an example, it has been bisected into two sub-polytopes by a line passing through its Chebyshev center.

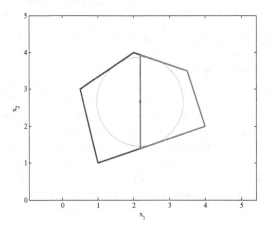

Fig. 3. The bisection of a polytope

4   Nonlinear ODE Reachability of Polytopes

In this section, the main contribution of the presented paper is described.

4.1   Principle

The procedures of the proposed approach involve geometric operations on polytopes such as the bisection of a polytope, the representation of a polytope by the intersection of zonotopes, the intersection of polytopes as well as the convex hull for a union of polytopes. The nonlinear ODE reachability with a polytope as the initial state can be computed by the following three steps:

– Represent the resulting polytope $\mathcal{P}_i$ at each step exactly by the intersection of m zonotopes: $Z_1, \cdots, Z_m$;
– Compute the evolution of these m zonotopes $Z_1, \cdots, Z_m$ (if any integration fails, either bisect the relevant zonotope or return to the previous step to bisect the polytope);
– Compute the intersection of these m solutions of the ODE at time T so as to obtain the renewed polytope $\mathcal{P}_{i+1}$;
– Return to the first step;

If the bisection of a zonotope or polytope is needed, the convex hull for the union of the bisected individual solutions is used instead to update the polytope $\mathcal{P}_i$ at each step. To guarantee that the algorithm terminates, under a given threshold on the volume of the zonotopes, the bisection is not performed, and the reachability cannot be computed.

4.2 Examples

In this section, two examples are presented to show the results of our approach.

Circle. The first example is the circle system. This latter is well-known and often used to demonstrate the robustness of an integration method to the wrapping effect.

Problem and Initial Conditions:
The problem is defined by the following equation:

$$\begin{cases} \dot{y}_1 = -y_2 \\ \dot{y}_2 = y_1 \end{cases} \tag{11}$$

The initial condition is taken in a polytope given by the five vertices $(-1, -3)$, $(-1.5, 3)$, $(0, 6)$, $(1.5, 4)$, and $(1, -4)$. Covered by the three zonotopes:

$$Z_1 = \begin{bmatrix} 0.1957 \\ 1.1522 \end{bmatrix} \oplus \begin{bmatrix} -1.6087 & -0.4130 \\ 0.8043 & 4.9565 \end{bmatrix} B^2, \ Z_2 = \begin{bmatrix} 0.25 \\ 0.5 \end{bmatrix} \oplus \begin{bmatrix} -1.8 & 1.55 \\ 2.4 & 3.1 \end{bmatrix},$$

$$\text{and } Z_3 = \begin{bmatrix} -0.1538 \\ 1.0385 \end{bmatrix} \oplus \begin{bmatrix} -1.3558 & 0.2981 \\ 1.8077 & 4.7692 \end{bmatrix} \ (\text{Fig. 4}).$$

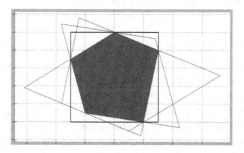

Fig. 4. Initial conditions of circle problem

Integration. The problem consists on computing the set of solution of the IVP at time $t_f = 50$. The RK4 method is used, with a tolerance of $1e{-}10$. The IVP is solved with the three previous zonotopes as initial conditions (Fig. 5). The IVP is also solved starting from the hull (box) of the polytope, because this is a common approach. The results are: the polytope obtained by the intersection of the solutions (zonotopes), the zonotope computed from the hull and the intersection of these two solutions (Fig. 6).

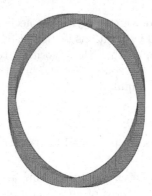

Fig. 5. Circle solution from $t = 0$ to $t = 50$ given in the form of a list of boxes.

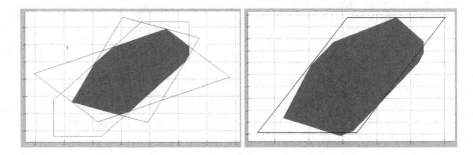

Fig. 6. Circle at $t = 50$: in black the three zonotopes, in red the polytope (left); and in black the zonotope obtained by the integration from the hull of the initial polytope, compared to the polytope (right). (Color figure online)

Discussion. In order to discuss the results, the volumes are computed and given in Table 1. The results show that our approach is efficient and better than the hull approach in terms of a smaller volume. It is also apparent that our solver is stable and robust against wrapping effect (important in the circle problem). This claim can be deduced by the fact that the volume of final set is close to the initial one. Finally, the intersection of the solution obtained from the polytope and the one obtained from the hull gives even a sharper result.

Volterra. The second example is the Volterra system. This latter is well-known and often used to demonstrate the efficiency of an integration method (alternation of contractive and dispersive parts).

Problem and Initial Conditions. The problem is defined by the following equation:

$$\begin{cases} \dot{y}_1 = 2y_1(1 - y_2) \\ \dot{y}_2 = -y_2(1 - y_1) \end{cases} \tag{12}$$

Table 1. Results in term of volume of the circle problem.

Initial polytope (IP)	21.25
Initial hull (IH)	30
Polytope (P) from IP	22.5046
Zonotope (Z) from IH	27.9348
Intersection of P and Z	21.8109

The initial condition is taken in a polytope given by the eight vertices $(1.1035, 3.0457)$, $(1.1041, 3.0386)$, $(1.0981, 3.0366)$, $(1.1039, 3.0358)$, $(1.0983, 3.0339)$, $(1.1020, 3.0320)$, $(1.0989, 3.0498)$ and $(1.0995, 3.0510)$. Covered by the three zonotopes:

$$Z_1 = \begin{bmatrix} 1.1007 \\ 3.0422 \end{bmatrix} \oplus \begin{bmatrix} -0.0032 & -0.0008 \\ 0.0016 & 0.0099 \end{bmatrix} B^2, \; Z_2 = \begin{bmatrix} 1.1000 \\ 3.0400 \end{bmatrix} \oplus \begin{bmatrix} -0.0036 & 0.0031 \\ 0.0048 & 0.0062 \end{bmatrix},$$

and $Z_3 = \begin{bmatrix} 1.1012 \\ 3.0395 \end{bmatrix} \oplus \begin{bmatrix} -0.0027 & 0.0006 \\ 0.0036 & 0.0095 \end{bmatrix}$ (Fig. 7).

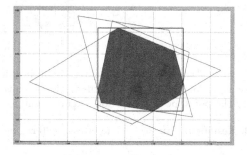

Fig. 7. Initial conditions of volterra problem

Integration. The problem consists on computing the set of solution of the IVP at time $t_f = 6$. The experimentation is the same than for the previous example (solution in Figs. 8 and 9).

Discussion. As for the first example, the volumes are computed and given in Table 2.

For this problem, as for the previous one, the polytopic approach is better in term of volume, than the zonotopic approach. However, there is also an interest to compute the solution with the both manner and to intersect the obtained results.

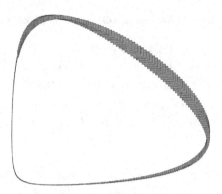

Fig. 8. Volterra solution

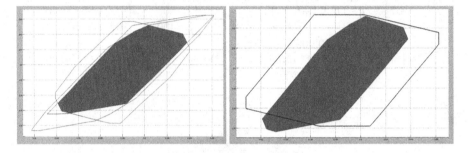

Fig. 9. Volterra problem at $t = 6$: in black the three zonotopes, in red the polytope (left); and in black the zonotope obtained by the integration from the hull of the initial polytope, compared to the polytope (right). (Color figure online)

Table 2. Results in term of volume of the Volterra problem.

Initial polytope (IP)	8.2505e−05
Initial hull (IH)	1.14e−04
Polytope (P) from IP	3.2273e−04
Zonotope (Z) from IH	5.8018e−04
Intersection of P and Z	3.0337e−04

5   Conclusion

In this paper, an approach to compute the reachability of nonlinear ODE with a polytopic set as the initial set is presented for the first time. Our method is based on the zonotopic representation of polytopes and on the zonotopic Runge-Kutta validated method. It has been shown through two well-known examples that our approach is efficient and robust with reduced wrapping effect. The volumes of the computed sets have been computed so as to validate this claim. The presented method is interesting in the field of hybrid systems, for example to compute the reachable sets as a linear program with polytopes. This method is then highly promising as a new approach for hybrid systems reachability. Moreover, the polytopic based Runge-Kutta allows one to propagate a set of inequalities (where the solution is a polytope) through a differential equation so as to obtain a novel set of inequalities (defining the polytope solution).

Acknowledgement. The second author was grateful for the financial support from the Engineering and Physical Sciences Research Council (EPSRC) of the U.K. under Grant EP/R005532/1.

References

1. Alexandre dit Sandretto, J., Chapoutot, A.: Validated explicit and implicit Runge-Kutta methods. Reliab. Comput. Electron. Ed. **22**, 78–108 (2016)
2. Bouissou, O., Chapoutot, A., Djoudi, A.: Enclosing temporal evolution of dynamical systems using numerical methods. In: Brat, G., Rungta, N., Venet, A. (eds.) NFM 2013. LNCS, vol. 7871, pp. 108–123. Springer, Heidelberg (2013). https://doi.org/10.1007/978-3-642-38088-4_8
3. Bouissou, O., Goubault, E., Putot, S., Tekkal, K., Vedrine, F.: HybridFluctuat: a static analyzer of numerical programs within a continuous environment. In: Bouajjani, A., Maler, O. (eds.) CAV 2009. LNCS, vol. 5643, pp. 620–626. Springer, Heidelberg (2009). https://doi.org/10.1007/978-3-642-02658-4_46
4. Butcher, J.C.: Coefficients for the study of Runge-Kutta integration processes. J. Aust. Math. Soc. **3**, 185–201 (1963)
5. de Figueiredo, L.H., Stolfi, J.: Self-validated numerical methods and applications. In: Brazilian Mathematics Colloquium Monographs, IMPA/CNPq (1997)
6. Eggers, A., Ramdani, N., Nedialkov, N., Fränzle, M.: Improving SAT modulo ODE for hybrid systems analysis by combining different enclosure methods. In: Barthe, G., Pardo, A., Schneider, G. (eds.) SEFM 2011. LNCS, vol. 7041, pp. 172–187. Springer, Heidelberg (2011). https://doi.org/10.1007/978-3-642-24690-6_13
7. Fukuda, K.: From the zonotope construction to the minkowski addition of convex polytopes. J. Symb. Comput. **38**(4), 1261–1272 (2004). Symbolic Computation in Algebra and Geometry
8. Hairer, E., Norsett, S.P., Wanner, G.: Solving Ordinary Differential Equations I: Nonstiff Problems, 2nd edn. Springer, Heidelberg (2009). https://doi.org/10.1007/978-3-540-78862-1
9. Henzinger, T.A., Horowitz, B., Majumdar, R., Wong-Toi, H.: Beyond HyTech: hybrid systems analysis using interval numerical methods. In: Lynch, N., Krogh, B.H. (eds.) HSCC 2000. LNCS, vol. 1790, pp. 130–144. Springer, Heidelberg (2000). https://doi.org/10.1007/3-540-46430-1_14
10. Kvasnica, M., Grieder, P., Baotić, M.: Multi-Parametric Toolbox (MPT) (2004)
11. Lohner, R.J.: Enclosing the solutions of ordinary initial and boundary value problems. In: Computer Arithmetic, pp. 255–286 (1987)
12. Moore, R.: Interval Analysis. Prentice Hall, Upper Saddle River (1966)
13. Nedialkov, N.S., Jackson, K., Corliss, G.: Validated solutions of initial value problems for ordinary differential equations. Appl. Math. Comp. **105**(1), 21–68 (1999)
14. Rump, S.M., Kashiwagi, M.: Implementation and improvements of affine arithmetic. In: Nonlinear Theory Applications, IEICE, vol. 6, no. 3, pp. 341–359 (2015)

15. Tucker, W.: A rigorous ode solver and smale's 14th problem. Found. Comput. Math. **2**(1), 53–117 (2002)
16. Wan, J., Sharma, S., Sutton, R.: Guaranteed state estimation for nonlinear discrete-time systems via indirectly implemented polytopic set computation. IEEE Trans. Autom. Control (2019). https://doi.org/10.1109/TAC.2018.2816262

The Satisfiability of Word Equations: Decidable and Undecidable Theories

Joel D. Day[1]([✉]), Vijay Ganesh[2], Paul He[2], Florin Manea[1], and Dirk Nowotka[1]

[1] Kiel University, Kiel, Germany
{jda,flm,dn}@informatik.uni-kiel.de
[2] University of Waterloo, Waterloo, Canada
vijay.ganesh@uwaterloo.ca, paul.he@edu.uwaterloo.ca

Abstract. The study of word equations is a central topic in mathematics and theoretical computer science. Recently, the question of whether a given word equation, augmented with various constraints/extensions, has a solution has gained critical importance in the context of string SMT solvers for security analysis. We consider the decidability of this question in several natural variants and thus shed light on the boundary between decidability and undecidability for many fragments of the first order theory of word equations and their extensions. In particular, we show that when extended with several natural predicates on words, the existential fragment becomes undecidable. On the other hand, the positive Σ_2 fragment is decidable, and in the case that at most one terminal symbol appears in the equations, remains so even when length constraints are added. Moreover, if negation is allowed, it is possible to model arbitrary equations with length constraints using only equations containing a single terminal symbol and length constraints. Finally, we show that deciding whether solutions exist for a restricted class of equations, augmented with many of the predicates leading to undecidability in the general case, is possible in non-deterministic polynomial time.

Keywords: Word equations · Decidability · Satisfiability

1 Introduction

A *word equation* is a formal equality $U = V$, where U and V are words (called the left and right side of the equation respectively) over an alphabet $A \cup X$; $A = \{\mathsf{a}, \mathsf{b}, \mathsf{c}, \ldots\}$ is the alphabet of *constants* or *terminals* and $X = \{x_1, x_2, x_3, \ldots\}$ is the set of *variables*. A *solution* to the equation $U = V$ is a morphism $h : (A \cup X)^* \to A^*$ that acts as the identity on A and satisfies $h(U) = h(V)$; h is called the assignment to the variables of the equation. For instance, $U = x_1 \mathsf{ab} x_2$ and $V = \mathsf{a} x_1 x_2 \mathsf{b}$ define the equation $x_1 \mathsf{ab} x_2 = \mathsf{a} x_1 x_2 \mathsf{b}$, whose solutions are the morphisms h with $h(x_1) = \mathsf{a}^k$, for $k \geq 0$, and $h(x_2) = \mathsf{b}^\ell$, for $\ell \geq 0$. An equation is *satisfiable* (in A^*) if it admits a solution $h : (A \cup X)^* \to A^*$. A set (or system) of equations is satisfiable if there exists an assignment of the variables of the

I. Potapov and P.-A. Reynier (Eds.): RP 2018, LNCS 11123, pp. 15–29, 2018.
https://doi.org/10.1007/978-3-030-00250-3_2

equations in this set that is a solution for all equations. In logical terms, word equations are often investigated as fragments of the first order theory $\mathsf{FO}(A^*, \cdot)$ of strings. Karhumäki et al. [18] showed that deciding the satisfiability of a system of word equations, that is, checking the truth of formulas from the existential theory Σ_1 of $\mathsf{FO}(A^*, \cdot)$, can be reduced to deciding the satisfiability of a single (more complex) word equation that encodes the respective system.

The existential theory of word equations has been studied for decades in mathematics and theoretical computer science with a particular focus on the decidability of the satisfiability of logical formulae defined over word equations. Quine [28] proved in 1946 that the first-order theory of word equations is equivalent to the first-order theory of arithmetic, which is known to be undecidable. In order to solve Hilbert's tenth problem in the negative [14], Markov later showed a reduction from word equations to Diophantine equations (see [21,22] and the references therein), in the hopes that word equations would prove to be undecidable. However, Makanin [22] proved in 1977 that the satisfiability of word equations *is* in fact decidable. Though Markov's approach was unsuccessful, similar ones, based on extended theories of word equations, can also be explored. Matiyasevich [25] showed in 1968 a reduction from the more powerful theory of word equations with linear length constraints (i.e., linear relations between word lengths) to Diophantine equations. Whether this theory is decidable remains a major open problem. More than a decade after Makanin's decidability result, the focus shifted towards identifying the complexity of solving word equations. Plandowski [27] showed in 1999 that this problem is in PSPACE. Recently, in a series of papers (see specifically e.g., [15,16]), Jez applied a new technique called recompression to word equations. This lead to, ultimately, a proof that the satisfiability of word equations can be decided in linear space. However, there is a mismatch between this upper bound and the known lower bound: solving word equations is NP-hard, but whether the problem is NP-complete remains open.

In recent years, deciding the satisfiability of systems of word equations has also become an important problem in fields such as formal verification and security where string solvers such as HAMPI [19], CVC4 [3], Stranger [31], ABC [2], Norn [1], S3P [29] and Z3str3 [4] have become more popular. However, in practice more functionality than just word equations is required, so solvers often extend the theory of word equations with certain functions (e.g., linear arithmetic over the length, replace-all, extract, reverse, etc.) and predicates (e.g., numeric-string conversion predicate, regular-expression membership, etc.). Most of these extensions are not expressible by word equations, in the sense introduced by Karhumäki et al. [18], and some of them lead to undecidable theories. On the one hand, regular (or rational) constraints or constraints based on involutions (allowing to model the mirror image, or, when working with equations in free groups, inverse elements), are not expressible, see [6,18], but adding them to word equations preserves the decidability [8]. As mentioned above, whether the theory of word equations enhanced with a length function is decidable is still a major open problem. On the other hand, the satisfiability of word equations extended with a *replace-all* operator was shown to be undecidable in [20], and the same holds when a numeric-string conversion predicate is added. Due to this very complex and fuzzy picture, none of the solvers mentioned above has a complete algorithm.

Our Contributions: In this setting, our work aims to provide a better understanding of the boundary between extensions and restrictions of the theory of word equations for which satisfiability is decidable and, respectively, undecidable.

Firstly, we present a series of undecidability results for the Σ_1-fragment of FO$(A^*, \cdot)$ extended with simple predicates or functions. In the main result on this topic, we show that extending Σ_1 with constraints imposing that a string is the morphic image of another one also leads to an undecidable theory. These results are related to the study of theories of quantifier-free word equations constrained by very simple relations, see, e.g., [6,13]. While our results do not settle the decidability of the theory of word equations with length constraints, they enforce the intuitive idea that enhancing the theory of word equations with predicates providing very little control on the combinatorial structure of the solutions of the equation leads to undecidability.

We further explore the border between decidability and undecidability when considering formulae over word equations allowing at most one quantifier alternation. We show that checking the truth of an arbitrary Σ_2-formula is equivalent to, on the one hand, checking the truth of a $\exists^*\forall^*$-quantified terminal-free formula, or, on the other hand, to a single $\exists^*\forall^*$-quantified inequation whose sides contain at most two terminals. Since the Inclusion of Pattern Languages problem (see [5,11,17]) can be reformulated as checking the truth of a single $\exists^*\forall^*$-quantified inequation whose sides contain at most two terminals and are variable disjoint, and it is undecidable, we obtain a clear image of the simplest undecidable classes of Σ_2-formulae. Consequently, we consider decidable cases. Complementary to the above, we show that the satisfiability in an arbitrary free monoid A^* of quantifier free *positive* formulae over word equations (formulae obtained by iteratively applying only conjunction and disjunction to word equations of the form $U = V$), in which we have at most one terminal $a \in A$ (appearing zero or several times) and no restriction on the usage of variables, enhanced with linear length constraints, is decidable, and, moreover, NP-complete. The decidability is preserved when considering positive Σ_2-formulae of this kind, as opposed to the case of arbitrary Σ_2 terminal-free formulae, mentioned above. Moreover, if we allow negated equations in the quantifier-free formulae (so arbitrary Σ_1-formulae) with at most one terminal, and length constraints, we obtain a decidable theory if and only if the general theory of equations with length constraints is decidable. Putting together these results, we draw a rather precise border between the decidable and undecidable subclasses of the Σ_2-fragment over word equations, defined by restrictions on the number of terminals allowed to occur in the equations and the presence or absence of inequations. As a corollary, we can show that deciding the truth of arbitrary formulae from the positive Σ_2-fragment of FO$(A^*, \cdot)$ (i.e., $\exists^*\forall^*$ quantified positive formulae), without length constraints, is decidable. The resulting proof follows arguments partly related to those in [9,23]. This result is strongly related to the work of [10,12,28], in which it was shown that the validity of sentences from the positive Π_2-fragment of $\mathsf{FO}(A^*, \cdot)$ (i.e., where the quantifier alternation was $\forall^* \exists^*$) is undecidable, as well as to the results of [30] in which it was shown that the truth of arbitrarily quantified positive formulae over word equations is decidable *over an infinite alphabet of terminals*.

We then extend our approach in a way partly motivated by the practical aspects of solving word equations. Most equations that can be successfully solved by string solvers (e.g., Z3str3) must be in *solved form* [12], or must not contain *overlapping variables* [32]. In a sense, this suggests that in practice it is interesting to find equations with restricted form that can be solved in reasonable time. We analyse, from a theoretical point of view, one of the simplest classes of equations that are not in solved form or contain equations with overlapping variables, namely strictly regular-ordered equations (each variable occurs exactly once in each side, and the order in which the variables occur is the same). We show that the satisfiability of such equations, even when enhanced with various predicates, is decidable. In particular we show that when extended with regular constraints (given by DFAs), linear length constraints, abelian equivalence constraints (two variables should be substituted for abelian-equivalent words), subword constraints (one variable should be a (scattered) subword of another), and Eq_a constraints (two variables should have the same number of occurrences of a letter a), the satisfiability problem remains NP-complete. Thus, there is hope that they can be solved reasonably fast by string solvers based on, e.g., SAT-solvers. This line of results is also related to the investigations initiated in [7,24], in which the authors were interested in the complexity of solving equations of restricted form. In the most significant result of [7], it was shown that deciding the satisfiability of strictly regular-ordered equations (with or without regular constraints) is NP-complete, which makes this class of word equations one of simplest known classes of word equations that are hard to solve. Although these results regard a very restricted class of equations, they might provide some insights in tackling harder classes, such as, e.g., quadratic equations.

The organization of the paper is as follows. In Sect. 2 we introduce the basic notions we use. In Sect. 3, we present firstly the undecidability results related to theories over word equations extended with various simple predicates, secondly the undecidability and decidability results related to quantifier alternation, and thirdly, we present the results related to strictly regular-ordered equations. Due to space constraints, some proofs are omitted, or only briefly sketched.

2   Preliminaries

Let $\mathbb{N}$ be the set of natural numbers, and let $\mathbb{N}_{\leq n}$ be the set $\{1, 2, \ldots, n\}$. Let A be an alphabet of letters (or symbols). Let A^* be the set of all words over A and ε be the empty word. Note that A^* is a monoid w.r.t. the concatenation of words. Let $|w|$ denote the length of a word w and for each $a \in A$, let $|w|_a$ denote the number of occurrences of a in w. For $1 \leq i \leq |w|$ we denote by $w[i]$ the letter on the ith position of w. A word w is *p-periodic* for $p \in \mathbb{N}$ (p is called a *period* of w) if $w[i] = w[i + p]$ for all $1 \leq i \leq |w| - p$; the smallest period of a word is called its *period*. If $w = v_1 v_2 v_3$ for some words $v_1, v_2, v_3 \in A^*$, then v_1 is called a *prefix* of w, v_1, v_2, v_3 are *factors* of w, and v_3 is a *suffix* of w. Two words w and u are called *conjugate* if there exist non-empty words v_1, v_2 such that $w = v_1 v_2$ and $u = v_2 v_1$. A word $v \in A^*$ is a *subword* of $w \in A^*$ if $v = v_1 \ldots v_k$, with $v_i \in A^*$, and $w = u_0 v_1 u_1 \cdots v_k u_k$, with $u_i \in A^*$. A word $z \in A^*$ is in the *shuffle* of $u, v \in A^*$, denoted $z \in u \Delta v$, if $z = u_1 v_1 \cdots u_k v_k$, with $u_i, v_i \in A^*$, and $u = u_1 \cdots u_k$, $v = v_1 \cdots v_k$. Two words $u, v \in A^*$ are *abelian equivalent* if $|u|_a = |v|_a$, for all $a \in A$. The following lemma is well known (see, e.g., [21]).

Lemma 1 (Commutativity Equation). *Let $v_1, v_2 \in A^*$. Then $v_1 v_2 = v_2 v_1$ if and only if there exist $w \in A^*$ and $p, q \in \mathbb{N}_0$ such that $v_1 = w^p$ and $v_2 = w^q$.*

Let $A = \{\mathsf{a}, \mathsf{b}, \mathsf{c}, \ldots\}$ be a finite alphabet of *constants* and let $X = \{x_1, x_2, \ldots\}$ be an alphabet of *variables*. Note that we assume X and A are disjoint, and unless stated otherwise, that $|A| \geq 2$. A word $\alpha \in (A \cup X)^*$ is usually called a *pattern*. For a pattern α and a letter $z \in A \cup X$, let $|\alpha|_z$ denote the number of occurrences of z in α; $\text{var}(\alpha)$ denotes the set of variables from X occurring in α. A morphism $h : (A \cup X)^* \to A^*$ with $h(a) = a$ for every $a \in A$ is called a *substitution*. A morphism $h : A^* \to B^*$ is a *projection* if $h(a) \in \{\epsilon, a\}$ for all $a \in A$. We say that $\alpha \in (A \cup X)^*$ is *regular* if, for every $x \in \text{var}(\alpha)$, we have $|\alpha|_x = 1$; e.g., $\mathsf{a} x_1 \mathsf{a} x_2 \mathsf{c} x_3 x_4 \mathsf{b}$ is regular. Note that $L(\alpha) = \{h(\alpha) \mid h \text{ is a substitution}\}$ (the pattern language of α) is regular when α is regular.

A (positive) *word equation* is a tuple $(U, V) \in (A \cup X)^* \times (A \cup X)^*$; we usually denote such an equation by $U = V$, where U is the left hand side (LHS, for short) and V the right hand side (RHS) of the equation. A negative word equation, or inequation, is the negation of a word equation, i.e., $\neg(U = V)$ or $U \neq V$.

A *solution* to an equation $U = V$ (resp., $U \neq V$), over an alphabet A, is a substitution h mapping the variables of UV to words from A^* such that $h(U) = h(V)$ (respectively, $h(U) \neq h(V)$); $h(U)$ is called the *solution word*. Note that we might ask whether a positive or negative equation has a solution over an alphabet larger than the alphabet of terminals that actually occur in the respective equation. A word equation is *satisfiable* over A if it has a solution over A, and the *satisfiability problem* is to decide for a given word equation whether it is satisfiable over a given alphabet A.

Karhumäki et al. [18] have shown that, given two equations E and E', one can construct the equations E_1, E_2, and E_3 that are satisfiable in A^*, with $|A| \geq 2$, if and only if $E \wedge E'$, $E \vee E'$, $\neg E$ are satisfiable respectively in A^*. In this construction, E_1 contains exactly the variables of E and E', while in E_2 and E_3 new variables are added with respect to those in the given equations; in all cases, even if E and E' were terminal-free, the new equations contain terminals. We use this result to show that for every quantifier-free first order formula over word equations we can construct a single equation that may contain extra variables and terminals, and is satisfiable if and only if the initial formula was satisfiable. Moreover, the values the variables of the initial equations may take in the satisfying assignments of the new equation are exactly the same values they took in the satisfying assignments of the initial formula. We also use in several occasions the following result from [18].

Lemma 2. *Let* $U, V, U', V' \in (X \cup A)^*$, $Z_1 = U\mathsf{a}U'U\mathsf{b}U'$, $Z_2 = V\mathsf{a}V'V\mathsf{b}V'$. *For any substitution* $h : X^* \to A^*$, $h(Z_1) = h(Z_2)$ *iff* $h(U) = h(V)$ *and* $h(U') = h(V')$.

In this paper we address equations with restricted form. A word equation $U = V$ is regular if both U and V are regular patterns. We call a regular equation *ordered* if the order in which the variables occur in both sides of the equation is the same; that is, if x and y are variables occurring both in U and V, then x occurs before y in U if and only if x occurs before y in V. Moreover, we say a regular-ordered equation is *strict* if each variable occurs in both sides. For instance $x_1\mathsf{a}x_2x_3\mathsf{b} = x_1\mathsf{a}x_2\mathsf{b}x_3$ is strictly regular-ordered while $x_1\mathsf{a} = x_1x_2$ is regular-ordered (but not strictly since x_2 occurs only on one side) and $x_1\mathsf{a}x_3x_2\mathsf{b} = x_1\mathsf{a}x_2\mathsf{b}x_3$ is regular but not regular-ordered.

In Sect. 3.3 we also consider equations with regular and linear length constraints defined as follows. Given a word equation $U = V$, a set of *linear length constraints* is a system θ of linear Diophantine equations where the unknowns correspond to the lengths of possible substitutions of each variable $x \in X$. Moreover, given a variable $x \in X$, a *regular constraint* is, in this paper, a regular language L_x given by a finite automaton; more general types of regular constraints, imposing that the image of a variable belongs to more than one language, are sometimes used (see [8] and the references therein). The satisfiability of word equations with linear length and/or regular constraints is the question of whether a solution h exists satisfying the system θ and/or such that $h(x) \in L_x$ for each $x \in X$.

3   Results

3.1   Undecidability Results

In this section, we show the undecidability of various extensions of the existential theory of word equations, defined as binary and 3-ary relations which may easily be interpreted as predicates. In each case, undecidability is ultimately obtained by showing that, for a unary-style encoding of integers following [6] (where a number is represented using the length of a string in the form $\mathsf{a}^*\mathsf{b}$, so ε is 0, b is 1, etc.), the additional predicate(s) can be used to define a multiplication predicate Multiply(x, y, z) which decides for numbers i, j, k encoded in this way (i.e., $x = \mathsf{a}^{i-1}\mathsf{b}, y = \mathsf{a}^{j-1}\mathsf{b}, z = \mathsf{a}^{k-1}\mathsf{b}$), whether $k = ij$. Since a corresponding addition predicate can easily be modelled for this encoding using only word equations, undecidability follows immediately.

Definition 1. *Let* AbelianEq, MorphIm, Projection, Subword $\subset A^* \times A^*$ *and* Shuffle, Insert, Erase $\subset A^* \times A^* \times A^*$ *be the relations given by:*

- $(x, y) \in$ AbelianEq *iff* x *and* y *are abelian-equivalent,*

- $(x, y) \in$ MorphIm *iff there exists a morphism* $h: A^* \to A^*$ *such that* $h(x) = y$,
- $(x, y) \in$ Projection *iff there exists a projection* $\pi: A^* \to A^*$ *such that* $\pi(x) = y$,
- $(x, y) \in$ Subword *iff* x *is a (scattered) subword of* y.
- $(x, y, z) \in$ Shuffle *iff* $z \in x \Delta y$,
- $(x, y, z) \in$ Erase *iff* z *is obtained from* x *by removing some occurrences of* y,
- $(x, y, z) \in$ Insert *iff* z *is obtained from* x *by inserting some occurrences of* y.

For each of the above relations we can also define a predicate with the same name which returns true iff the tuple of arguments belongs to the relation.

The membership problems for all the above relations are in NP, and therefore decidable. Our main result of this section concerns the MorphIm predicate:

Theorem 1. *Let* $|A| \geq 3$. *Then given the predicate* MorphIm, *the predicate* Multiply *is definable by an existential formula.*

Proof. Assume that A contains at least three distinct letters: $\mathsf{a}, \mathsf{b}, \mathsf{c}$. We shall actually define a predicate $\text{Multiply}_2(x, y, z)$ which returns true iff $x = \mathsf{a}^i\mathsf{b}, y = \mathsf{a}^j\mathsf{b}, z = \mathsf{a}^{ij}\mathsf{b}$ and $ij \geq 2$. Note that we can immediately obtain Multiply from this, as $\text{Multiply}(x, y, z) = \text{Multiply}_2(\mathsf{a}x, \mathsf{a}y, \mathsf{a}z)$ for $x, y, z \neq \varepsilon$ (assuming also $x = y = z = \mathsf{b}$ does not hold). The exceptional cases, when $ij < 2$ can easily be handled individually. We define first a predicate checking some 'initial conditions':

$$init(x, x', x'', y, y', z, z') := \exists w, w', w''.x' = w\mathsf{a} \wedge y' = w'\mathsf{a} \wedge (x' = w''\mathsf{a}\mathsf{a} \vee y' = w''\mathsf{a}\mathsf{a})$$
$$\wedge x'\mathsf{a} = \mathsf{a}x' \wedge y'\mathsf{a} = \mathsf{a}y' \wedge z'\mathsf{a} = \mathsf{a}z' \wedge x = x'\mathsf{b} \wedge y = y'\mathsf{b} \wedge z = z'\mathsf{b} \wedge x''x = xx''.$$

Recalling Lemma 1, it is straightforward to see that $init$ evaluates to true if and only if there exist $i, j, k, \ell, p \in \mathbb{N}_0$ with $ij \geq 2$ such that (1) $x' = \mathsf{a}^i$, $y' = \mathsf{a}^j$, $z' = \mathsf{a}^k$, and (2) $x = \mathsf{a}^i\mathsf{b}$, $y = \mathsf{a}^j\mathsf{b}$, $z = \mathsf{a}^k\mathsf{b}$, and (3) $x'' = (\mathsf{a}^i\mathsf{b})^p$. Now we give the definition of Multiply_2 as follows:

$$\text{Multiply}_2(x, y, z) := \exists x', x'', y', z', u, v. \, init(x, x', x'', y, y', z, z') \wedge \text{MorphIm}(x'', y')$$
$$\wedge \text{MorphIm}(y', x'') \wedge \text{MorphIm}(u, v) \wedge u = x''\mathsf{cc}x''x'\mathsf{ccb} \wedge v = z'\mathsf{cc}z'x'\mathsf{cc}.$$

Suppose that Conditions (1)–(3) are met (i.e., $init$ is satisfied). Consider the subclause $\text{MorphIm}(x'', y') \wedge \text{MorphIm}(y', x'')$. This is satisfied if and only if there exist morphisms $g, h : A^* \to A^*$ such that $g((\mathsf{a}^i\mathsf{b})^p) = \mathsf{a}^j$ and $h(\mathsf{a}^j) = (\mathsf{a}^i\mathsf{b})^p$. Clearly, the latter implies that p is a multiple of j, while the former implies that j is a multiple of p, and hence if both are satisfied then $j = p$. On the other hand, if $j = p$, then it is easy to construct such morphisms (g maps b to a and a to ε while h maps a to $\mathsf{a}^i\mathsf{b}$). Thus this subclause is satisfied in addition to the $init$ predicate if and only if Conditions (1)–(3) hold for $p = j$. By elementary substitutions, the remaining part (i.e., $\text{MorphIm}(u, v) \wedge u = x''\mathsf{cc}x''x'\mathsf{ccb} \wedge v = z'\mathsf{cc}z'x'\mathsf{cc}$) is also satisfied if and only if $u = (\mathsf{a}^i\mathsf{b})^j\mathsf{cc}(\mathsf{a}^i\mathsf{b})^j\mathsf{a}^i\mathsf{ccb}$, and $v = (\mathsf{a}^k\mathsf{cc}\mathsf{a}^{k+i}\mathsf{cc})$. It remains to show that there exists a morphism $f : A^* \to A^*$ such that $f(u) = v$ if and only if $k = ij$. In the case that $k = ij$, the morphism f may be given e.g. by $f(\mathsf{a}) = \mathsf{a}$, $f(\mathsf{b}) = \varepsilon$ and $f(\mathsf{c}) = \mathsf{c}$. For the other direction, assume that such a morphism f exists. Firstly, consider the case that $f(c) \in \{a, b\}^*$. Then c must occur in $f(a)$ or $f(b)$. However, under our assumption that $ij \geq 2$, this implies $|f(u)|_c > 4$ meaning $f(u) \neq v$ which is a contradiction. Consequently, we may infer that $f(c)$ contains the letter c. Then since $|u|_c = |v|_c$, it follows that $f(c) = v_1 c v_2$ where $v_1, v_2 \in \{a, b\}^*$. Thus $f(u) = f((a^i b)^j) v_1 c v_2 v_1 c v_2 f((a^i b)^j) a^i v_1 c v_2 v_1 c v_2 f(b)$. It follows that $v_1 = v_2 = \varepsilon$, and thus that $f(b) = \varepsilon$. Hence, $f(a^{ij}) = a^k$ and $f(a^{ij+i}) = a^{k+i}$ must hold. Clearly, $f(a) = a^n$ for some $n \in \mathbb{N}$. Thus we have $nij = k$ and $nij + ni = k + i$. Hence, $n = 1$ and $k = ij$, as required. $\qquad\square$

Unlike for the other predicates below, our construction for MorphIm relies strictly on the alphabet A having at least three letters. This is in particular contrast to many other results on the (un)decidability theories of word equations which are usually independent of alphabet size (provided $|A| \neq 1$). Thus we consider it to be of particular interest to settle the remaining open case of whether Theorem 1 holds also for binary alphabets A.

As previously mentioned, further to the predicate MorphIm, many other natural predicates dealing with basic properties and relationships of words lead to undecidability. The following result concerns the remaining predicates listed in Definition 1, and it is obtained by reducing to predicates $Onlyas(x, y)$ and $Onlybs(x, y)$ which return true if and only if $y = a^{|x|_a}$ (respectively, $y = b^{|x|_b}$). Büchi and Senger [6] show how these predicates can easily be used to model multiplication, and thus undecidability follows.

Proposition 1. *Given any of the predicates* AbelianEq, Shuffle, Projection, Subword, Insert, Erase, *the predicates* Onlyas *and* Onlybs *are definable by existential formulas.*

The next theorem sums up the consequences of Proposition 1 and Theorem 1.

Theorem 2. *The existential theory of word equations becomes undecidable when augmented with any of the following predicates:* AbelianEq, Shuffle, Projection, Subword, MorphIm *(if $|A| \geq 3$),* Insert, Erase.

3.2   Quantifier Alternation

Next, we focus on extending the existential theory of word equations by allowing, instead of new predicates, quantifier alternation.

Firstly, recall the *Inclusion of Pattern Languages* problem (IPL, for short, see [5,17]): given two patterns $\alpha \in (A \cup X)^*$ and $\beta \in (A \cup Y)^*$, where A is an alphabet of constants with at least two distinct letters and X and Y are disjoint sets of variables, decide whether $L(\alpha) \subseteq L(\beta)$. IPL admits a reformulation in terms of word equations: decide whether the formula $\exists x_1, \ldots, x_n. \forall y_1, \ldots, y_m. \alpha \neq \beta$ holds in A^*. As IPL is undecidable for terminal alphabets of size 2 or more [5,11], it immediately follows that checking the truth value of $\exists^* \forall^*$-quantified inequation $U \neq V$ in A^*, with $|A| \geq 2$, is undecidable even when

U and V do not contain any common variable, as long as the number of terminals occurring in UV is at least two. This exhibits a very simple fragment of Σ_2 that is undecidable.

Further, we show two normal form results for the Σ_2-fragment of $FO(A^*, \cdot)$.

Proposition 2. *Let $A \neq \emptyset$ be an alphabet. For every formula ϕ in the Σ_2-fragment of $FO(A^*, \cdot)$ we can construct a Σ_2 terminal-free formula ψ, which holds in A^* iff ϕ holds in A^*.*

Proposition 3. *Let A be an alphabet, $|A| \geq 2$. For every formula ϕ in the Σ_2-fragment of $FO(A^*, \cdot)$ we can construct $\psi = \exists x_1, \ldots, x_n . \forall y_1, \ldots, y_m . U \neq V$, with $U, V \in (A \cup \{x_1, \ldots, x_n, y_1, \ldots, y_m\})^*$, such that ϕ holds in A^* if and only if ψ holds in A^*.*

Note that Proposition 3 does not follow directly by applying the results of [18] to the initial arbitrary formula, in order to reduce it to a single equation. This would have lead to an $\exists^* \forall^* \exists^*$-quantified positive equation, so not to a Σ_2-formula.

The results in Propositions 2 and 3 as well as the remarks regarding IPL show that it is undecidable to check whether some very simple formulae hold in A^*, when $|A| \geq 2$. Also, it is worth noting that applying first Proposition 2 and then Proposition 3 to an arbitrary Σ_2-formula would lead to a single $\exists^* \forall^*$-quantified inequation which contains two terminals, as the constructions in [18] (used in the proof of Proposition 3) require at least two terminals in the equation. However, unlike the inequations encoding IPL instances, the one we obtain by applying our two propositions does not necessarily fulfil the condition that its sides are variable disjoint. Thus, it is natural to ask whether every Σ_2-formula can be reduced to an inequation encoding an instance of IPL. We conjecture that the answer to this question is no.

We have showed that deciding whether a Σ_2-formula, whose sides contain two terminals, holds in A^* for some $|A| \geq 2$ is undecidable. It is possible to show that, when $|A| \geq 2$, for every word equation (which can encode any formula from the Σ_1-fragment of $FO(A^*, \cdot)$, by [18]) we can construct a word equation whose sides contain exactly two terminals a and b, and whose solutions over $\{\mathsf{a}, \mathsf{b}\}$ bijectively correspond to the solutions of the initial equation. Thus, solving a word equation whose sides contain two terminals is as complex as solving arbitrary word equations.

Hence, we will investigate next which is the case of Σ_1 and Σ_2-formulae over word equations whose sides contain at most one terminal. Proposition 2 already gives us a first answer: checking whether a Σ_2-terminal-free formula holds in A^*, with $|A| \geq 2$, is undecidable. On the other hand, checking whether a formula from $FO(A^*, \cdot)$, whose sides contain at most one terminal a, holds in $\{\mathsf{a}\}^*$ is decidable, as it can be canonically seen as a formula in the Presburger arithmetic.

We concentrate now on other decidable variants. In all these cases, we augment our signature with linear arithmetic over the lengths of variables; all decidability results obtained in this setting hold canonically for the case when such restrictions do not appear. We first look at equations without any quantifier alternation.

Proposition 4. *Let* a $\in A$. *The satisfiability in* A^* *of quantifier-free positive formulae over word equations* $U = V$, *with* $U, V \in (X \cup \{a\})^*$, *with linear length constraints is NP-complete.*

Complementing the above result, we show that the satisfiability of quantifier-free first order formulae over word equations $U = V$ (so including negation), such that $U, V \in (X \cup \{a\})^*$, with linear length constraints is equivalent to solving arbitrary word equations with length constraints. Hence, at the moment, we cannot say anything about the decidability of such formulae. One direction of our result is immediate, while the other follows similarly to Proposition 2.

Theorem 3. *Let* $|A| \geq 2$ *and* a $\in A$. *Given an equation* $U = V$, *with* $U, V \in (A \cup X)^*$, *with linear length constraints* θ, *there exists a system* S *of positive and negative equations* $U_i = V_i$ *or* $U_i \neq V_i$ *with* $U_i, V_i \in (X' \cup \{a\})^*$ *and* $X \subset X'$ *with linear length constraints* θ', *such that* S *is satisfiable (in* A^* *) if and only if* $U = V$ *is satisfiable.*

Building on Proposition 4, Theorem 4 considers the Σ_2 fragment in the case that only one terminal letter may appear in the equations. Note that this does not necessarily imply $|A| = 1$. If the positive theory only is considered, augmented with the Length predicate defined in the previous section (i.e., Length(x, y) is true if and only if $|x| = |y|$), then we obtain a decidable fragment. Note in particular that the Length predicate can be used in conjunction with simple equations to model arbitrary linear length constraints.

Theorem 4. *Let* a $\in A$. *The positive* Σ_2*-fragment, restricted to word equations containing only the terminal symbol* a, *augmented with* Length, *is decidable.*

Firstly, we need the following lemma. Then, we give the full proof of Theorem 4.

Lemma 3. *Let* $Y = \{y_1, y_2, \ldots, y_n\} \subseteq X$ *and let* $U, V \in (Y \cup A)^*$. *Let* $k > |UV|$ *and let* $h : X^* \to A^*$ *be the substitution such that* $h(y_i) = ab^{k+i}a$. *Then* $h(U) = h(V)$ *if and only if* $U = V$ *(the strings* U *and* V *coincide).*

Proof. (Theorem 4) W.l.o.g. we may assume that all arguments of the Length predicate are either single variables or words in A^*. Indeed, if we have a "longer" argument α over $(X \cup A)^*$, we can replace it with a new variable x and add the equation $x = \alpha$. For the purposes of this proof we shall say that a term is trivial if, for all the word equations $U = V$, U and V are identical, and moreover, all Length predicates of the form Length(x, y) where either $x = y \in X$ or $x, y \in A^*$ and $|x| = |y|$. If $|A| = 1$, decidability follows from the decidability of Presburger arithmetic. Thus we may assume a, b $\in A$ with a $\neq$ b. W.l.o.g. we may assume that we have a sentence in disjunctive normal form as follows:

$$\exists x_1, x_2, \ldots, x_n. \forall y_1, y_2, \ldots, y_m. (e_{1,1} \wedge \ldots \wedge e_{1,k_1}) \vee \ldots \vee (e_{t,1} \wedge \ldots \wedge e_{t,k_t}), \quad (1)$$

where the $e_{i,j}$ are either: (1) of the form Length(z_1, z_2) where z_1 and z_2 are in $\{x_1, \ldots, x_n, y_1, \ldots, y_m\} \cup A^*$, or (2) individual word equations over the variables $x_1, \ldots, x_n, y_1, \ldots, y_m$ and the terminal a.

We shall show that an assignment for $x_1, x_2, \ldots, x_n$ satisfies (1) if and only if there exists $s, 1 \le s \le t$ such that all the resulting atoms $e_{s,i}$ become trivial. The 'if' direction is straightforward, thus we consider the 'only if' direction. Suppose the $x_1, x_2, \ldots, x_n$ are fixed, and consider the result of each $e_{i,j}$ under the substitution. Suppose that for each $s, 1 \le s \le t$ there exists $r_s, 1 \le r_s \le k_s$ such that e_{s,r_s} is non-trivial. Let p be the maximum over the lengths of all constant terms in the sentence, lengths of the x_i, and lengths of equations given by the type-(2) atoms $e_{i,j}$ for $1 \le i \le t, 1 \le j \le k_i$. Consider the choice of $y_1, y_2, \ldots, y_m$ given by $y_k = \mathsf{ab}^{p+k}\mathsf{a}$ for $1 \le k \le m$. By Lemma 3, if e_{s,r_s} is of type (2), then it will evaluate to false. If e_{s,r_s} is of type (1), then we have three cases. Firstly, if both arguments to the Length predicate are constant terms in A^*, then clearly e_{s,r_s} will evaluate to false since it is non-trivial. Similarly, since the y_i are longer than all constant terms and substituted values of the x_ks, if exactly one of the arguments is a constant in A^* while the other is a variable in $\{y_1, y_2, \ldots, y_m\}$, then e_{s,r_s} will also evaluate to false. Finally, since $|y_\ell| \ne |y'_\ell|$ for all $\ell \ne \ell'$, if both arguments are variables, e_{s,r_s} will again evaluate to false. Summarising the above, for any given choice of $x_1, x_2, \ldots, x_n$ there exists a choice of $y_1, y_2, \ldots, y_m$ such that any of the conjunctions containing a non-trivial equation or Length predicate will be false. It follows that the sentence is satisfiable if and only if there exists a choice for $x_1, x_2, \ldots, x_n$ and $s, 1 \le s \le t$ such that all the $e_{s,i}$ terms, $1 \le i \le k_s$ become trivial.

For terms $e_{i,j}$ of type (2), this is reduced to solving a system of existentially quantified word equations over $x_1, x_2, \ldots, x_n$ as follows: suppose $e_{i,j}$ is the equation $u_0 y_{i_1} u_1 y_{i_2} u_2 \ldots y_{i_p} u_p = v_0 y_{j_1} v_1 y_{j_2} \ldots y_{j_q} v_q$, where $p, q \in \mathbb{N}_0$, $i_k, j_\ell \in [1, m]$ for $1 \le k \le p$ and $1 \le \ell \le q$, and $u_k, v_\ell \in (\{x_1, x_2, \ldots, x_n\} \cup A)^*$ for $1 \le k \le p$ and $1 \le \ell \le q$. Clearly, for a given choice of values for $x_1, \ldots, x_n$, the equation $e_{i,j}$ becomes trivial if and only if $p = q$, and $u_0 = v_0, u_1 = v_1, \ldots, u_p = v_p$, that is, if $x_1, \ldots, x_n$ forms a solution to the system of equations $u_0 = v_0, u_1 = v_1, \ldots, u_p = v_p$ over the variables $x_1, \ldots, x_n$ and terminal symbols from A.

For a term $e_{i,j}$ of type (1), observe that they may only become trivial under some substitution for the x_ℓs either if it is already trivial, in which case it can just be removed, or if both arguments are in $\{x_1, x_2, \ldots, x_n\}$. Thus, any of the clauses $(e_{i,1} \wedge \ldots \wedge e_{i,k_i})$ containing a term $e_{i,j}$ not conforming to these two cases can be removed entirely. After these two phases of removal, it remains to solve, for each $s, 1 \le s \le t$, a system of equations (i.e., the conjunctions of the systems derived from the $e_{s,j}$ terms of type (2), as described above) subject to a system of linear length constraints (derived from the terms of type (1)). The resulting equations will also only contain the terminal symbol a, since they are taken directly from the original equations, so the decidability follows from Proposition 4. □

Note that the reasoning above can be modified in a straightforward way to get decidability of the positive Σ_2 fragment in the general case (but without length constraints), by substituting any of the well-known algorithms for solving existentially quantified systems of equations (e.g. Makanin's algorithm, Plandowski's algorithm, Recompression) in place of Proposition 4. The resulting proof has similar arguments to those of [9,23], although these results do not address this case directly. Also, the decidability result shown in Theorem 4 is, in a sense, optimal, as checking the truth of terminal-free arbitrary Σ_2-formulae is undecidable.

Corollary 1. *The truth of Σ_2^+-formulae over A^* is decidable.*

3.3   Decidability with Restricted Form

Following the results of the previous section, we explore one more decidable fragment of $FO(A^*, \cdot)$. More precisely, instead of restricting the terminal symbols appearing in the equation(s) we restrict the variables, considering one of the simplest cases of equations that are not in solved form, thus right at the border of the equations that can be solved by practical string solvers [12,32]. We are able to obtain decidability when augmenting the theory simultaneously with linear arithmetic over variable lengths, regular constraints given as DFAs, as well as constraints based on the predicates Subword and Eq_a from the previous section. Formally, we say that subword (resp. Eq_a, abelian) constraints are sets of pairs of variables $(x, y) \in X^2$. Solving equation with these constraints requires asserting that for each such pair, the corresponding predicate returns true (so that, e.g., for each abelian constraint (x, y), the substitutions for x and y are abelian equivalent).

Theorem 5. *The problem of solving strictly regular ordered equations with regular constraints given by DFAs, linear length constraints, Eq_a constraints (for each $a \in A$), abelian constraints, and subword constraints is NP-complete.*

Proof. Here we present only a sketch of the proof of this theorem. The proof rests on the fact that solutions to strictly regular ordered equations have a particularly well-suited form for parameterisation. In particular, by applying some canonical arguments from the field of combinatorics on words, it can be shown that the set of solutions is spanned by parametric solutions of the form $h(x) = (u_x v_x)^{n_x} u_x$ where $|u_x v_x|$ is linear in the length of the equation, and n_x may be any positive integer if x is "overlapping" (i.e., some part of $h(x)$ on the LHS coincides with part of $h(x)$ on the RHS) and 0 otherwise. Thus, when deciding if a solution exists to the equation which also satisfies the length and regular constraints, it is sufficient to firstly guess such a parametric form, and then decide whether there exist values for the parameters n_x such that the additional constraints are satisfied. Deciding which values of the parameters are also valid under the regular constraints can be done in an efficient way (non-deterministically) due to Lemma 4: simply guess them. Subword constraints are handled in the same way due to a similar technical result, Lemma 5.

Lemma 4. *Let L be a regular language given by a DFA, M, with n states. Let $u, v \in A^*$. Then there exist $q \in \mathbb{N}_{\leq n}$, $P, S \subseteq \mathbb{N}_{\leq n} \cup \{0\}$ such that the intersection of $(uv)^+ u$ and L is given by $\{(uv)^s u \mid s \in S\} \cup \{(uv)^{q\mu+p} u \mid \mu \in \mathbb{N} \wedge p \in P\}$.*

Lemma 5. *Let $u, v, u', v' \in A^*$. Let $S = \{(p, q) \mid (uv)^p u$ is a subword of $(u'v')^q u'\}$. Then either $S = \emptyset$, or there exist integers p_1, p_2, $q_1, q_2, q_{3,0}, \ldots, q_{3,p_2-1}$, and $r_1, r_2, \ldots, r_{p_1+p_2-1}$, all polynomial in $|uvu'v'|$, such that*

$$S = S' \cup \bigcup_{1 \leq i < p_2} \{(p, q) \mid p = p_1 + kp_2 + i \wedge q \geq q_1 + kq_2 + q_{3,i} \wedge k \in \mathbb{N}\}$$

where $S' = \{(p, q) \mid p < p_1 + p_2 \wedge q \geq r_p\}$. Moreover, this representation of S can be computed in (nondeterministic) polynomial time.

Having so-far obtained expressions for parametric solutions satisfying the equation and the regular constraints and subword constraints, it remains to check whether any of the remaining possibilities also satisfy the length, abelian, and Eq_a constraints. It is straightforward, having already guessed the values u_x and v_x, to convert the latter two into length constraints. Thus finding solutions satisfying all constraints is eventually reduced to solving a linear system of Diophantine equations where the unknowns are the parameters. Since the resulting coefficients can be shown to be at most exponentially large, this is possible in non-deterministic polynomial time, see [26].   □

For regular-ordered equations without the strictness (i.e. variables may occur in only one side), the equivalent of Theorem 5 does not hold. It is a straightforward exercise that regular-ordered equations where each side has only one singly-occurring variable, with regular constraints given by DFAs, is PSPACE-complete. This follows from the fact that determining whether the intersection of n DFAs is empty is PSPACE-hard. Similarly, the undecidability proofs for the predicates described in Sect. 3.1 require only very restricted combinations of equations, so we can expect that when such constraints are added, strict restrictions on the structure are necessary for maintaining decidability.

References

1. Abdulla, P.A., Atig, M.F., Chen, Y., Holík, L., Rezine, A., Rümmer, P., Stenman, J.: Norn: an SMT solver for string constraints. In: Kroening, D., Păsăreanu, C.S. (eds.) CAV 2015. LNCS, vol. 9206, pp. 462–469. Springer, Cham (2015). https://doi.org/10.1007/978-3-319-21690-4_29
2. Aydin, A., Bang, L., Bultan, T.: Automata-based model counting for string constraints. In: Kroening, D., Păsăreanu, C.S. (eds.) CAV 2015. LNCS, vol. 9206, pp. 255–272. Springer, Cham (2015). https://doi.org/10.1007/978-3-319-21690-4_15
3. Barrett, C., Conway, C.L., Deters, M., Hadarean, L., Jovanović, D., King, T., Reynolds, A., Tinelli, C.: CVC4. In: Gopalakrishnan, G., Qadeer, S. (eds.) CAV 2011. LNCS, vol. 6806, pp. 171–177. Springer, Heidelberg (2011). https://doi.org/10.1007/978-3-642-22110-1_14
4. Berzish, M., Ganesh, V., Zheng, Y.: Z3str3: a string solver with theory-aware heuristics. In: Proceedings of the FMCAD 2017, pp. 55–59. IEEE (2017)
5. Bremer, J., Freydenberger, D.D.: Inclusion problems for patterns with a bounded number of variables. Inf. Comput. **220**, 15–43 (2012)

6. Büchi, J.R., Senger, S.: Definability in the existential theory of concatenation and undecidable extensions of this theory. Z. für math. Logik Grundlagen d. Math. **47**, 337–342 (1988)
7. Day, J.D., Manea, F., Nowotka, D.: The hardness of solving simple word equations. In: Proceedings of the MFCS 2017. LIPIcs, vol. 83, pp. 18:1–18:14 (2017)
8. Diekert, V., Jeż, A., Plandowski, W.: Finding all solutions of equations in free groups and monoids with involution. Inf. Comput. **251**, 263–286 (2016)
9. Diekert, V., Lohrey, M.: Existential and positive theories of equations in graph products. Theory Comput. Syst. **37**(1), 133–156 (2004)
10. Durnev, V.G.: Undecidability of the positive ∀∃-theory of a free semigroup. Sib. Math. J. **36**(5), 917–929 (1995)
11. Freydenberger, D.D., Reidenbach, D.: Bad news on decision problems for patterns. Inf. Comput. **208**(1), 83–96 (2010)
12. Ganesh, V., Minnes, M., Solar-Lezama, A., Rinard, M.: Word equations with length constraints: what's decidable? In: Biere, A., Nahir, A., Vos, T. (eds.) HVC 2012. LNCS, vol. 7857, pp. 209–226. Springer, Heidelberg (2013). https://doi.org/10. 1007/978-3-642-39611-3_21
13. Halfon, S., Schnoebelen, P., Zetzsche, G.: Decidability, complexity, and expressiveness of first-order logic over the subword ordering. In: Proceedings of the LICS 2017, pp. 1–12. IEEE Computer Society (2017)
14. Hilbert, D.: Mathematische probleme. Nachrichten von der Gesellschaft der Wissenschaften zu Göttingen, Mathematisch-Physikalische Klasse **1900**, 253–297 (1900)
15. Jeż, A.: Recompression: a simple and powerful technique for word equations. In: Proceedings of the STACS 2013. LIPIcs, vol. 20, pp. 233–244 (2013)
16. Jeż, A.: Word equations in nondeterministic linear space. In: Proceedings of the ICALP 2017. LIPIcs, vol. 80, pp. 95:1–95:13 (2017)
17. Jiang, T., Salomaa, A., Salomaa, K., Yu, S.: Decision problems for patterns. J. Comput. Syst. Sci. **50**(1), 53–63 (1995)
18. Karhumäki, J., Mignosi, F., Plandowski, W.: The expressibility of languages and relations by word equations. J. ACM (JACM) **47**(3), 483–505 (2000)
19. Kiezun, A., Ganesh, V., Guo, P.J., Hooimeijer, P., Ernst, M.D.: HAMPI: a solver for string constraints. In: Proceedings of the ISSTA 2009, pp. 105–116. ACM (2009)
20. Lin, A.W., Barceló, P.: String solving with word equations and transducers: towards a logic for analysing mutation XSS. In: ACM SIGPLAN Notices. vol. 51, pp. 123–136. ACM (2016)
21. Lothaire, M.: Combinatorics on Words. Addison-Wesley, Boston (1983)
22. Makanin, G.S.: The problem of solvability of equations in a free semigroup. Sb.: Math. **32**(2), 129–198 (1977)
23. Makanin, G.S.: Decidability of the universal and positive theories of a free group. Math. USSR-Izv. **25**(1), 75 (1985)
24. Manea, F., Nowotka, D., Schmid, M.L.: On the solvability problem for restricted classes of word equations. In: Brlek, S., Reutenauer, C. (eds.) DLT 2016. LNCS, vol. 9840, pp. 306–318. Springer, Heidelberg (2016). https://doi.org/10.1007/978-3-662-53132-7_25
25. Matiyasevich, Y.V.: A connection between systems of words-and-lengths equations and hilbert's tenth problem. Zapiski Nauchnykh Seminarov POMI **8**, 132–144 (1968)
26. Papadimitriou, C.H.: On the complexity of integer programming. J. ACM (JACM) **28**(4), 765–768 (1981)

27. Plandowski, W.: Satisfiability of word equations with constants is in PSPACE. In: Proceedings of the FOCS 1999, pp. 495–500. IEEE (1999)
28. Quine, W.V.: Concatenation as a basis for arithmetic. J. Symb. Log. **11**(4), 105–114 (1946)
29. Trinh, M.-T., Chu, D.-H., Jaffar, J.: Progressive reasoning over recursively-defined strings. In: Chaudhuri, S., Farzan, A. (eds.) CAV 2016. LNCS, vol. 9779, pp. 218–240. Springer, Cham (2016). https://doi.org/10.1007/978-3-319-41528-4_12
30. Vazenin, J.M., Rozenblat, B.V.: Decidability of the positive theory of a free countably generated semigroup. Math. USSR Sb. **44**(1), 109–116 (1983)
31. Yu, F., Alkhalaf, M., Bultan, T.: STRANGER: an automata-based string analysis tool for PHP. In: Esparza, J., Majumdar, R. (eds.) TACAS 2010. LNCS, vol. 6015, pp. 154–157. Springer, Heidelberg (2010). https://doi.org/10.1007/978-3-642-12002-2_13
32. Zheng, Y., Ganesh, V., Subramanian, S., Tripp, O., Berzish, M., Dolby, J., Zhang, X.: Z3str2: an efficient solver for strings, regular expressions, and length constraints. Form. Methods Syst. Des. **50**(2–3), 249–288 (2017)

Left-Eigenvectors Are Certificates of the Orbit Problem

Steven de Oliveira[1,2]([⊠]), Virgile Prevosto[2], Peter Habermehl[3], and Saddek Bensalem[1]

[1] Université Grenoble Alpes, Grenoble, France
de.oliveira.steven@gmail.com
[2] CEA, List, Palaiseau, France
[3] IRIF, Université Paris Diderot - Paris 7, Paris, France

Abstract. This paper investigates the connection between the Kannan-Lipton Orbit Problem and the polynomial invariant generator algorithm PILA based on eigenvectors computation. Namely, we reduce the problem of generating linear and polynomial certificates of non-reachability for the Orbit Problem for linear transformations with coefficients in $\mathbb{Q}$ to the generalized eigenvector problem. Also, we prove the existence of such certificates for any transformation with integer coefficients, which is not the case with rational coefficients.

1 Introduction

Finding a suitable representation of the reachable set of configurations for a given transition system or transformation is a fundamental problem in computer science, notably in program analysis and verification. An exact representation of the reachable set can generally not be exactly computed. In this context, invariants often provide a good balance between precision, conciseness and ease of use. Model-checking [13] and deductive verification [9] often require the user to provide invariants in order to reach a given proof objective. In practice, for large programs, manually writing each invariant for each loop is extremely costly and becomes quickly infeasible. Users can rely on invariants synthesizers, that manage to infer an over-approximation of the reachable set of configurations. Abstract interpretation [1,3] for example is based on the propagation of abstract values, such as *e.g.* intervals or octagons, that encompass the whole set of possible concrete inputs. Dynamic inference [7] tries to infer a candidate invariant satisfied by a large amount of runtime executions. The quality of the synthesis is here dependent of the chosen invariant pattern. Mathematical properties of specific kinds of transformations, such as the use of linear algebra properties [2,4] or the search of algebraic dependencies [12] can elegantly facilitate the automated search for invariants. For all of these techniques, the following issues arise:

1. they work under very specific hypotheses;
2. generated invariants may not be precise enough to succeed in proving or *disproving* a given property.

© Springer Nature Switzerland AG 2018
I. Potapov and P.-A. Reynier (Eds.): RP 2018, LNCS 11123, pp. 30–44, 2018.
https://doi.org/10.1007/978-3-030-00250-3_3

As an example, [4,5] describe the PILA method for generating invariants of linear transformations based on the eigenspace problem. This method relies on the stability of left-eigenvectors of a linear transformation: a left-eigenvector φ of a linear transformation f verifies $\varphi \circ f = \lambda \varphi$ for some constant λ. Depending on the value of λ, φ leads to inductive invariants. For instance, if $\lambda = 1$, then $\forall X, \varphi \circ f(X) = \varphi(X)$, hence the relation $\varphi(X)$ remains constant. When $|\lambda| \leqslant 1$ (respectively $|\lambda| \geqslant 1$), then the PILA technique generates inductive invariants of the form $|\varphi(X)| \leqslant k$ (respectively $|\varphi(X)| \geqslant k$). All polynomial equality invariants $(P(X) = k)$ and some inequality invariants (every P such that $P(X) \leqslant k$ and a subset of P such that $P(X) \geqslant k$) can be generated with this technique. PILA has been developed in the context of polynomial invariant generation, an already widely studied topic [2,14]. One of the purposes of this article is to study the usefulness of such invariants for solving the Kannan-Lipton Orbit Problem.

The Kannan-Lipton Orbit Problem

A particular instance of the reachability problem is called the *Kannan-Lipton Orbit Problem* [10,11], which can be stated as follows:

Given a square matrix $A \in \mathcal{M}_d(\mathbb{Q})$ of size d and

two vectors $X, Y \in \mathbb{Q}^d$, determine if there exists n such that $A^n X = Y$.

This problem is decidable in polynomial time. In the case an instance of the problem has no solution (in other words, Y is not reachable from X), [8] studies the existence of non-reachability semialgebraic certificates for a given instance of the Orbit Problem where Y is not reachable. Semialgebraic certificates are sets described by conjunctions and disjunctions of polynomial inequalities with integer coefficients that include the reachable set of states but not the target Y. These certificates allow to quickly prove the non-reachability of the given vector Y and all vectors outside of the certificate. [8] concludes on the existence of such certificates under simple hypotheses on the eigenvalue decomposition of A.

These hypotheses are surprisingly similar to the hypotheses of PILA as, when $|\lambda| \neq 1$, left-eigenvectors represent polynomial inequality invariants while [8] uses certificates defined by *polynomial inequalities*. The PILA technique is sometimes unable to infer invariants, especially when the studied matrix is non-diagonalizable with all its eigenvalues λ such that $|\lambda| = 1$, while [8] is able to infer certificates. A slight extension of PILA presented in this article solves this problem by using *generalized eigenvectors* which we show can be used as certificates. Also, we shortly conclude on non diagonalizable matrices with eigenvalues λ such that $|\lambda| = 1$ and on matrices with integer coefficients. Depending on the cases presented in Table 1, we will prove that:

- in the first hypothesis, there exists a linear transformation of dimension $O(n^2)$ (resp. $O(2^n)$) computing an equivalent image of A such that its eigenvectors can be used as real certificates (resp. semialgebraic certificates) for the non reachability of the given instance;

Table 1. Comparaison between PILA, the results of [8] and the contributions of this paper

	Hypothesis 1	Hypothesis 2	Hypothesis 3
Hypotheses on matrix A with eigenvalue λ	$\|\lambda\| \neq 0 \wedge \|\lambda\| \neq 1$	A not diagonalizable $\|\lambda\| = 1$	A diagonalizable $\|\lambda\| = 1$
PILAT [4,5]	Inequality invariants $P(X) \leqslant 0, P(X) \geqslant 0$	Equality invariants $P(X) = 0$	Equality invariants $P(X) = 0$
[8] on the existence of certificates	General existence of a semialgebraic certificate	General existence of a semialgebraic certificate	Necessary & sufficient conditions for the existence of a semialgebraic certificate
Contributions	– Existence of M computing the same image as A – Eigenvectors of M are certificates	– Existence of M computing the same image as A – Generalized eigenvectors of M are certificates	– Eigenvectors can be used as certificates under the same conditions

- in the second hypothesis, there exists a linear transformation of dimension $O(n^2)$ (resp. $O(2^n)$) computing an equivalent image of A such that its *generalized* eigenvectors can be used as real certificates (resp. semialgebraic certificates) for the non reachability for the given instance;
- in a more general case, a semialgebraic certificate for the Orbit Problem in $\mathbb{Z}$ always exists.

It is worth noting that to our knowledge, there exists no proof about the decidability of the existence of linear certificates directly on A.

Interest of Eigenvectors. The Jordan Normal form of a matrix used in [8] can be calculated in polynomial time given eigenvectors and generalized eigenvectors. It is however necessary to compute *all eigenvectors and generalized eigenvectors* of a transformation to get the Jordan Normal form. Here, in most cases we only need the calculation of a subset of eigenvectors.

2   Setting

Let $\mathbb{K}$ be a field and $d \in \mathbb{N}$. Given two vectors u, v of same dimension, we note $\langle u, v \rangle = u^t.v$, with . the usual dot product (i.e. the sum of the product of each component of u and v). A linear combination of variables is defined by a single vector φ such that $v \to \langle \varphi, v \rangle$. Every linear transformation $f : \mathbb{K}^d \to \mathbb{K}^d$ corresponds to a square matrix $A_f \in \mathcal{M}_d(\mathbb{K})$. For any vector $\varphi \in \mathbb{K}^d$, $\varphi^t :$

$\mathbb{K}^d \rightarrow \mathbb{K}$ will denote a linear transformation. When the context is clear, we will refer to A_f as A. The transformation obtained by n successive applications of a transformation $f : \mathbb{K}^d \rightarrow \mathbb{K}^d$ is denoted by f^n and its matrix is A_f^n. Affine transformations can be considered as linear transformation by adding an extra dimension. For example, the transformation $f(x) = x + 1$ can be considered equivalent to the transformation $g(x, 1) = (x + 1, 1)$. In this way, every affine transformation also admits a unique matrix representation.

Definition 1. *Let $f : \mathbb{K}^d \rightarrow \mathbb{K}^d$ be a linear transformation and A its associated matrix. Then, $\varphi \in \mathbb{K}^d$ (respectively $\varphi \in \mathbb{K}^d \rightarrow \mathbb{K}$) is called a λ-right-eigenvector (resp. λ-left-eigenvector) and λ its corresponding eigenvalue if $A * \varphi = \lambda\varphi$ (resp. $\varphi^t * A = \lambda\varphi^t$).*

When a concept can be applied to either left or right-eigenvectors, we will simply refer to them as eigenvectors.

Definition 2. *A family of linked generalized λ-eigenvectors $\mathcal{F}_f = \{e_0, ..., e_k\}$ for the transformation f are vectors verifying for all $i \leqslant k, f(e_0) = \lambda e_0$ and $f(e_i) = \lambda e_i + e_{i-1}$.*

The Orbit Problem. This article focuses on $\mathbb{A} \subset \mathbb{C}$, the field of algebraic numbers. Elements of $\mathbb{A}$ are roots of polynomials with integer coefficients. Indeed, the linear transformations we consider are in $\mathbb{Q}^d \rightarrow \mathbb{Q}^d$, thus their eigenvalues (as roots of the characteristic polynomial) are in $\mathbb{A}$. Let $f : \mathbb{Q}^d \rightarrow \mathbb{Q}^d$ be a linear transformation. We refer to the Orbit Problem of A_f with an initial vector $X \in \mathbb{Q}^d$ and a target vector $Y \in \mathbb{Q}^d$ as $\mathcal{O}(A, X, Y)$. In other words, $\mathcal{O}(A, X, Y) = (\exists n \in \mathbb{N}.Y = A^n X)$.

Definition 3. *A non-reachability certificate or just certificate is a couple $(N, P) \in \mathbb{N} \times \mathcal{P}(\mathbb{Q}^d)$ of an instance $\mathcal{O}(A, X, Y)$ such that:*

- $\forall n \in \mathbb{N}, n < N \Rightarrow A^n X \neq Y$
- $\forall n \in \mathbb{N}, n \geqslant N \Rightarrow A^n X \in P$
- $Y \notin P$

N is called the certificate index *and P the* certificate set.

When the certificate set is described by conjunctions and disjunctions of linear (resp. polynomial) combinations of variables, the certificate is called linear (resp. polynomial). Irrational, semialgebraic and rational certificates are linear or polynomial certificates whose coefficients are respectively irrationals, algebraic integers or rationals.

Semi-algebraic certificates, are always equivalent to rational certificates. Indeed, every coefficient $\varphi_i \in \mathbb{A}$ is nullified by a polynomial Q with integer coefficients. It is then possible to replace φ_i by a free variable that is constrained to be a root of Q. For example, $P = \{x | \sqrt{2}x \leqslant 2\} = \{x | \exists y.y^2 = 2 \wedge y \geqslant 0 \wedge yx \leqslant 2\}$.

Remarks. This definition of certificates is slightly different than the notion of certificates of [8] as it does not require an inductivity criterion. We have chosen this notation so as to simplify the article.

The certificate sets we generate are *future invariants* of the transformation, in the sense that $f^n(X)$ eventually reaches the set for some n and always remains in it, whereas Y is outside the invariant. Different choices of X and Y may delay the number of iterations needed to reach it. The certificate index solves this issue by expressing the number of iterations necessary for $f^n(X)$ to reach the certificate set. This information is crucial for the practical use of certificates, as a solver can use it to shorten its analysis.

The existence of such a couple implies the non reachability of Y as $A^n X$ is either different from Y or belongs to a set to which Y does not. For example, if Y does not belong to the reachable set of states $R = \{A^n X \mid n \geqslant 0\}$, the couple $(0, R)$ is a certificate. However, typically, R can not be described in a *non-enumerative* way. We are interested in *simple* certificates, i.e. where proving that the objective Y does not belong to the reachable set of states is straightforward. That means that membership in P should be easy to solve. For example, let $R' = \{(v_1, ..., v_n) \in \mathbb{Q}^n : v_1 + v_2 \geqslant 0\}$ and assume $R \subset R'$. Testing whether Y is in R' or not is easy as this set is described by a linear combination of variables. If $Y \notin R'$, then R' is generally a *better* (simpler) certificate set than R. On the other hand, finding a good certificate index may be harder. Its search is studied in Sect. 3.1.

3   Invariants by Generalized Eigenvectors

3.1   Certificate Sets of the Rational Orbit Problem

The decidability of the existence or the non-existence of semialgebraic certificates for the Orbit Problem for rational linear transformations is proven in [8]. It classifies four categories of rational matrices A:

- A admits null eigenvalues;
- A has at least an eigenvalue of modulus strictly greater or less than 1;
- A has all its eigenvalues of modulus 1, but it is not diagonalisable;
- A has all its eigenvalue of modulus 1 and is diagonalisable.

In the second case, linear transformations always admit a non reachability certificate if the Orbit problem has no solution. The intuition behind this result is to consider the Jordan normal form J of the matrix A. Let V be a vector of variables and V_J the vector of variables in the base of J. In this form, for any eigenvalue λ of A, there exists a variable v_J (representing a linear combination of variables of V) such that $J * V_{J|_{v_J}} = \lambda v_J$. Applied k times, the new value of v_J is $\lambda^k v_J$, which diverges towards infinity or converges towards 0 when $|\lambda| \neq 1$. Checking if a value y is reachable or not can then be done by checking if there exists $k \in \mathbb{N}$ such that $\lambda^k v_J = y$. We are now left to compute those certificates.

Case 1: There Exist Null Eigenvalues
This particular case leads to degenerate instances of the orbit problem. When a linear transformation admits a null eigenvalue, there exists a linear combination of variables that is always null. In other words, there exists a variable v that

can be expressed as a linear combination of the other variables. Therefore, this variable doesn't provide any useful information on the transformation other than an easily checkable constraint on v. If the linear constraint is satisfied, we get rid of this case by using Lemma 6 of [8], stating the following:

Lemma 1. *The problem of generating non-reachability certificates for an orbit instance $\mathcal{O}(A, X, Y)$ can be reduced to the problem of generating reachability certificates for an orbit instance $\mathcal{O}(A', X', Y')$ where A' is invertible.*

Case 2: There Exist Eigenvalues λ and $|\lambda| \neq 1$.

Real Eigenvalues. The key of the following property lies in [5], stating that λ-left eigenvectors φ of a linear transformation A are its invariants. More precisely, we can see that if φ is a left-eigenvector of A, then by definition the following holds:

$$\forall v \in \mathbb{K}^d, \langle \varphi, Av \rangle = \lambda \langle \varphi, v \rangle \tag{1}$$

If $|\lambda| > 1$ (resp. $|\lambda| < 1$), then the sequence $(|\langle \varphi, A^n v \rangle|)$ (for $n \in \mathbb{N}$) is *strictly increasing* (resp. *strictly decreasing*),

Property 1. *Let $A \in \mathcal{M}_d(\mathbb{Q})$ a linear transformation and $\mathcal{O}(A, X, Y)$ an instance of the Orbit problem with no solution. Searching for a non-reachability certificate of an instance of the Orbit problem when A admits real eigenvalues λ such that $|\lambda| \neq 0$ and $|\lambda| \neq 1$ can be reduced to computing the eigenvector decomposition of A.*

More precisely, if there exists φ a λ-left-eigenvector of A with $|\lambda| \neq 0$ and $|\lambda| \neq 1$, then the couple (N, P) defined as follows is a non-reachability certificate of $\mathcal{O}(A, X, Y)$.

1. *If $|\langle \varphi, X \rangle| \neq 0$ and $|\langle \varphi, Y \rangle| = 0$, then $N = 0$ and $P = \{v : \langle \varphi, v \rangle \neq 0\}$*
2. *If $|\langle \varphi, X \rangle| = 0$ and $|\langle \varphi, Y \rangle| \neq 0$, then $N = 0$ and $P = \{v : \langle \varphi, v \rangle = 0\}$.*
3. *If $|\langle \varphi, X \rangle| \neq 0$ and $|\langle \varphi, Y \rangle| \neq 0$, $N = max(1, \lfloor \frac{ln(|\langle \varphi, Y \rangle|) - ln(|\langle \varphi, X \rangle|)}{ln(|\lambda|)} \rfloor + 1)$ and*
   - *If $|\lambda| > 1$, then $P = \{v : |\langle \varphi, v \rangle| \geq |\lambda . \langle \varphi, Y \rangle|\}$.*
   - *If $|\lambda| < 1$, then $P = \{v : |\langle \varphi, v \rangle| \leq |\lambda . \langle \varphi, Y \rangle|\}$.*
4. *Otherwise, if $d > 1$ there exist a transformation $B \in \mathcal{M}_{d-1}(\mathbb{Q})$ such that the problem of finding a certificate for $\mathcal{O}(A, X, Y)$ can be reduced to the problem of finding a certificate for $\mathcal{O}(B, X', Y')$ with X' and $Y' \in \mathbb{Q}^{d-1}$. If $d = 1$, then $\mathcal{O}(A, X, Y)$ has a solution.*

The certificate is linear iff $\lambda \in \mathbb{Q}$.

Proof. Let φ be a left-eigenvector of A associated to the eigenvalue λ. We know that for all v, $\langle \varphi, v \rangle = k \Rightarrow \langle \varphi, Av \rangle = \lambda . k$. Let $U_n = |\langle \varphi, A^n X \rangle|$ be the n-th reachable state from X. If $|\lambda| < 1$ (resp. $|\lambda| > 1$), then (U_n) is strictly decreasing (resp. strictly increasing).

1. Let $k_v = |\langle \varphi, v \rangle|$. If $k_X \neq 0$ and $k_Y = 0$, then the sequence (U_n) never reaches k_Y, as for all n, $U_n \neq 0$. In other words, $|U_n| > 0$ for all $n \in \mathbb{N}$. Then it is clear that $P = \{X : |\langle \varphi, X \rangle| \neq 0\}$ is a valid certificate set of index $N = 0$.

2. Similarly, if $k_X = 0$ and $k_Y \neq 0$, then $P = \{X : |\langle \varphi, X \rangle| = 0\}$ and $N = 0$.

3. Assume now that $k_X \neq 0$ and $k_Y \neq 0$. If $k_X < k_Y$ and $|\lambda| < 1$ (respectively $k_X > k_Y$ and $|\lambda| > 1$), then $(1, \{v : |\langle \varphi, v \rangle| \leqslant |\lambda|.k_Y\})$ is a valid certificate set (respectively $(1, \{v : |\langle \varphi, v \rangle| \geqslant |\lambda|.k_Y\}))$. Otherwise, let us assume $|\lambda| < 1$ and $k_X \geq k_Y$. U_n is strictly decreasing, thus there exists a N such that $U_N \geqslant k_Y$ and $U_{N+1} < k_Y$. This implies that Y can only be reachable after a finite number of iterations N. We also have that $U_{N+1} \geqslant |\lambda|.k_Y$ and $U_{N+2} < |\lambda|.k_Y$. If for all $n \leq N$, $Y \neq A^n X$, we can define $P = \{v \mid \langle \varphi, v \rangle| < |\lambda|.k_Y\}$, and obtain $Y \notin P$ and $\{A^{N+1+n}X \mid n\mathbb{N}\} \subset P$. Therefore, the couple $(N+1, P)$ is a non-reachability certificate of $\mathcal{O}(A, X, Y)$. A similar proof for $|\lambda| > 1$ is valid as the sequence U_n is now strictly increasing and the couple $(N, \{|\langle \varphi, X \rangle| \geqslant |\lambda|.k_Y\})$ is the corresponding certificate. We will now study the exact value of N. If Y is reachable, then there exists a unique value of N such that $|\lambda|^N|\langle \varphi, X \rangle| = k_Y$. This value is precisely $\frac{ln(|\langle \varphi, Y \rangle|) - ln(|\langle \varphi, X \rangle|)}{ln(|\lambda|)}$. If for every value of $n \leqslant N$, Y is not reached and as Y does not belong to the certificate set P, the couple $(max(1, \lfloor N \rfloor + 1), P)$ is a non-reachability certificate.

4. Assume $k_X = k_Y = 0$. In this case for every n, $\langle \varphi, A^n X \rangle = 0$. There exists a base $\mathcal{B}$ of the transformation in which there exists a variable v which remains null for every iteration of the transformation. In other words, there exist A', Q such that $A' = Q.A.Q^{-1}$.

   Assume $d > 1$ and let $B' = A'_{|V \setminus v}$ and $Q' = Q_{|V \setminus v}$ the transformations restricted to all variables but v (by removing both the associated line and column). Finding a certificate for A is reduced to finding a certificate for $B = Q'^{-1}B'Q'$.

   If $d = 1$ and there exist a linear combination φ of X such that $\langle \varphi, X \rangle = 0$, then $X = 0$. Similarly, $Y = 0$.

Concerning the linearity of the certificate, if $\lambda \in \mathbb{Q}$, then every coefficient of φ also belongs to $\mathbb{Q}$. Indeed A has rational coefficients, so does $\varphi A = \lambda.\varphi$. Similarly, if φ has rational coefficients, $\varphi.A = \lambda.\varphi$ also does.

In the case of $k_X \neq 0$ and $k_Y \neq 0$, we also have to get rid of the absolute value around $\langle \varphi, v \rangle$ in the definition of the certificate set. If $|\lambda| > 1$, the certificate set $\{v : (\langle \varphi, v \rangle \geqslant |\lambda\langle \varphi, Y \rangle|) \wedge (\langle \varphi, v \rangle \leqslant -|\langle \varphi, Y \rangle|)\}$ is linear. A similar set can be found for $|\lambda| < 1$.   □

Certificate Index. Being able to minimize the number of necessary unrollings to prove the non reachability is useful. In this regard, notice that the certificate index value N of Property 1 is such that for every $n < N$, $\langle \varphi, A^n X \rangle \notin P$. In other words, it is minimal for its associated certificate set.

Example. Consider the Orbit Problem $\mathcal{O}(A, X, Y)$ with

$$A = \begin{pmatrix} 0 & 3 & 0 & 0 \\ -3 & 3 & 1 & 0 \\ 0 & 0 & 2 & 1 \\ 1 & 1 & 0 & 1 \end{pmatrix}$$

A admits two real eigenvalues $\lambda_1 \approx 0.642$ and $\lambda_2 \approx 2.48$ respectively associated to the left-eigenvectors $\varphi_1 = (-0.522, 0.355, -0.261, 0.73)$ and $\varphi_2 = (0.231, -0.36, -0.749, -0.506)$. This is enough to build two preliminary certificate sets that only depend on Y : $P_1 = \{v.|\langle\varphi_1, v\rangle| \leqslant \lambda_1.|\langle\varphi_1, Y\rangle|\}$ and $P_2 = \{v.|\langle\varphi_2, v\rangle| \geqslant \lambda_2.|\langle\varphi_2, Y\rangle|\}$. Those can be used for any initial valuation of X.

Let's now set $X = (1, 1, 1, 1)$ and $Y = (-9, -7, 28, 7)$. We have then

- $\langle\varphi_1, X\rangle = 0.302$ and $\langle\varphi_1, Y\rangle = 0.015$, so $N = 7$.
- $\langle\varphi_2, X\rangle = -1.384$ and $\langle\varphi_2, Y\rangle = -24.073$, so $N = 4$.

We can easily verify that for any $n \leqslant 7$, $A^n X \neq Y$, so the certificates $(7, P_1)$ and $(4, P_2)$ are sufficient to prove the non reachability of Y.

Complex Eigenvalues. The treatment of complex eigenvalues can be reduced to the Case 1 by the *elevation* method described in [4]. The idea is simple: if variables evolves linearly (or affinely) then any monomial of those variables also evolves linearly (or affinely). For example, given $f(x) = x + 1$, then the new value of x^2 after application of f is $(x + 1)^2 = x^2 + 2x + 1$, which is an affine combination of x^2, x and 1. f can be *elevated* to the degree 2 by expressing this new monomial: $f_2(x_2, x) = (x_2 + 2x + 1, x + 1)$.

Definition 4. *Let $A \in \mathcal{M}_d(\mathbb{K})$. We denote $\Psi_k(A)$ the elevation matrix such that $\forall X \in \mathbb{K}^n, \Psi_k(A).p(X) = p(A.X)$, with $p \in (\mathbb{K}[X]^k)$ a polynomial associating X to all possible monomials of degree k or lower.*
By extension, we denote $\Psi_k(v)$ a vector v elevated to the degree k.

A and $\Psi_d(A)$ represents the same application, except that $\Psi_d(A)$ also calculates monomial values of variables manipulated by A. Hence, certificates of $\mathcal{O}(\Psi_d(A), \Psi_d(X), \Psi_d(Y))$ are also certificates for $\mathcal{O}(A, X, Y)$, We also have the following property [4]:

Property 2. *Let $A \in \mathcal{M}_d(\mathbb{Q}), \Lambda(M)$ the eigenvalue set of a matrix M and k an integer. Then for any product p of k or less elements of $\Lambda(A), p \in \Lambda(\Psi_k(A))$ where $\Psi_k(A)$ is the elevation of A to the degree k.*

The product of all eigenvalues is the determinant of the transformation, which is by construction a rational. The elevation to the degree n where n is the size of the matrix admits then at least one rational eigenvalue. We can deduce from this the following theorem.

Theorem 1. *Let $\mathcal{O}(A, X, Y)$ be an unsatisfiable instance of the Orbit problem with $A \in \mathcal{M}_n(\mathbb{Q})$ admitting at least one eigenvalue $\lambda \in \mathbb{C}$ such that $|\lambda| \neq 0$ and $|\lambda| \neq 1$. Then left eigenvectors of $\Psi_d(A)$ provide:*

- *real linear semialgebraic certificates for $d = 1$ ($\Psi_1(A) = A$) if there exist real eigenvalues;*
- *real semialgebraic certificates of degree 2 for $d = 2$ if there exist complex eigenvalues;*
- *at least one rational certificate of degree n for $d = n$ if $|det(A)| \neq 1$.*

Proof. We treat each case separately:

- The case where A admits real eigenvalues is treated by Property 1;
- If A admits a complex eigenvalue λ, A also admits its conjugate $\bar{\lambda}$ as eigenvalue. By Property 2, $\Psi_2(A)$ admits $\lambda.\bar{\lambda}$ as a real eigenvalue, which is treated by Property 1;
- The product of all eigenvalues of a rational matrix is rational. As such, Ψ_n necessarily admit a rational eigenvalue which implies the existence of an associated rational eigenvector that can be used, according to Property 1, as a certificate.

$\square$

Remark. The image of $A \in \mathcal{M}_d(\mathbb{K})$ is a projection of the image of $\Psi_k(A)$ for any k, and semialgebraic certificates of A are, by extension, semilinear certificates of $\Psi_n(A)$. The size of $\Psi_k(A)$ is $\binom{d+k}{k}$, which is $O(d^2)$ when $k = 2$ and $O(d^d)$ when $d = k$. An eigenvector computation has a polynomial time complexity (slightly better than $O(d^3)$). The two first cases of Theorem 1 are thus computable in polynomial time in the number of variables.

Example. The matrix from the previous example admits two complex eigenvalue $\lambda \approx 1.439 + 2.712i$ and $\bar{\lambda}$. As $\lambda\bar{\lambda} \approx 9.425$, it also admits a polynomial invariant φ (whose size is too long to fit in this article as it manipulates 10 monomials). However, $\langle \varphi, X \rangle = 0.220$ and $\langle \varphi, Y \rangle = 195.738$, thus the associated index is 4.

Case 3: All Eigenvalues Have a Modulus Equal to 1 and the Matrix is Not Diagonalisable

Real Eigenvalues. This case is trickier as eigenvectors do not give information about the convergence or the divergence of the linear combination of variables they represent. For example, let us study the orbit problem $\mathcal{O}(A, X, Y)$ where A is the matrix associated with the mapping $f(x, \mathbb{1}) = (x+2*\mathbb{1}, \mathbb{1})$, $X = (0, 1)$ and $Y = (5, 1)$. x_Y is odd, thus Y is not reachable. f admits only $\varphi = (0, 1)$ as left-eigenvector associated to the eigenvalue $\lambda = 1$, meaning that $\langle (0, 1), (x, \mathbb{1}) \rangle = \langle (0, 1), f(x, \mathbb{1}) \rangle$ for any x. As $\langle (0, 1), (x, \mathbb{1}) \rangle = \mathbb{1}$, we are left with the invariant $\mathbb{1} = 1$. This invariant is clearly insufficient to prove that Y is not reachable.

f thankfully admits a generalized left-eigenvector $\mu = (\frac{1}{2}, 1)$ associated to 1. More precisely, $\mu A = \mu + \varphi$, which implies that $\mu A^n X = (\mu + n\varphi).X$. In other words, we have $\frac{1}{2}x + 1 = \frac{1}{2}x_X + 1 + n$ which simplifies into $\frac{1}{2}x = n$. The couple $(3, \{(x, y) : \exists n > 3, \frac{1}{2}x = n\})$ is a non reachability certificate.

Property 3. *Let A be a non-diagonalisable linear transformation, X a vector and $\{e_i\}_{i<N}$ N linked 1-left eigenvectors[1] (i.e. $e_0 A = e_0$ and for $0 < i < N$, $e_i A = e_i + e_{i-1}$). For all $1 \geqslant i < N$, $\langle e_i A^k, X \rangle = P_i(k)$, where $P_i(k)$ is a polynomial of non null degree in the variable k if and only if there exist $j < i$ such that $\langle e_j, X \rangle \neq 0$.*

---

[1] The existence of such a family with $N > 1$ is guaranteed by the non diagonalisability of A.

Proof. Let $\{e_i\}_{i<N}$ a family of N linked 1-left eigenvectors. We can calculate $P_i(k)$ by induction on i. For $i = 1$, e_0 verifies $e_0 A^k = e_0 + k * e_1$. Hence, $\langle e_0 A^k, X \rangle = \langle e_0, X \rangle$ is a polynomial of non null degree iff $\langle e_0, X \rangle \neq 0$.

Assume now $e_i.A^k = P_i(k)$ is a vector of polynomials of non null degree. Then, we have $e_{i+1}.A^{k+1} = (e_{i+1} + e_i).A^k = e_{i+1}A^k + P_i(k)$ Now, let $U_{n+1} = U_k + P_i(n)$. Then for U_0, $U_k = U_0 + \sum_{l=0}^{k} P_i(l)$ is a vector of polynomials of non null degree. As well as in the case $i = 1$, $P_{i+1}(k)$ has a non null degree if and only if for all $j < i$, $\langle e_j, X \rangle \neq 0$ as every polynomial expression of $P_{i+1}(k)$ contains $\langle e_j, X \rangle$. □

As every polynomial eventually diverges, there exists a linear combination of variables of X that diverges if X follows the hypothesis of this property. Otherwise, [8] have shown in Lemma 6 that the existence of a certificate for such instances is equivalent to the existence of certificates that are treated in the Case 4. Indeed, expressing a matrix A in the Jordan Normal form is exactly expressing A in the base of eigenvectors. The hypothesis of Property 3 matches the third part of Lemma 6 from [8].

Remark. Even if the first eigenvector is enough to represent a non-reachability certificate, every generalized eigenvector also can. By Property 3, the value of the linear combination described by a generalized eigenvector φ evolves polynomially, thus it eventually always decrease or increase (after the highest root of its derivate). That is why for a given objective Y there exist a finite number of n such that $|\varphi Y| \leqslant |\varphi A^n X|$, thus after this n, $\{v : |\varphi v| > |\varphi Y|\}$ is a certificate.

Complex Eigenvalues. If $\lambda \in \mathbb{C}$, we will use the same trick we used for complex eigenvalues of Case 2. As for every complex eigenvalue λ of A, $\bar{\lambda}$ is also an eigenvalue, then $\lambda.\bar{\lambda} = 1$ is an eigenvalue of $\Psi_2(A)$ by property 2. Thus:

Theorem 2. *Let $\mathcal{O}(A, X, Y)$ be a non satisfiable instance of the Orbit Problem such that for all eigenvalue λ of A, $|\lambda| = 1$ and A is not diagonalisable. Then there exist a family of 1-left-eigenvectors $\mathcal{F} = \{e_0, ..., e_n\}$ of $\Psi_2(A)$ such that for all $1 \leqslant i \leqslant n$, $Q_i(n) = \langle e_i, \Psi_2(A)^n \Psi_2(X) \rangle$ is a non-constant polynomial if and only if there exist $j < i$ such that $\langle e_i, X \rangle \neq 0$ and (N, P) is a non reachability certificate with:*

- *$N = \lfloor max(\{0\} \cup \{x \in \mathbb{R}.Q_i(x) = \langle e_i, \Psi_2(A^x)\Psi_2(Y) \rangle\}) \rfloor$*
- *$P = \{v : |\langle e_i, \Psi_2(A)^n \Psi_2(v) \rangle| \geqslant |Q_i(N)|\}$*

Proof. Let $\mathcal{O}(A, X, Y)$ be an instance of the Orbit Problem. We will reduce the problem to the case where A has positive rational eigenvalues, i.e. $\lambda = 1$ and A admits a family $\mathcal{F}$ of left-eigenvectors of size $|\mathcal{F}| > 1$. In this case, by Property 3 we know that there exists a linear combination of variables v following a polynomial evolution described by Q such that $deg(Q) > 0$. As Q eventually diverges, there exists a N such that for all $N' > N$, $|v(A^{N'} X)| > |v(Y)|$. This N is the maximum between 0 and the highest value of x such that $Q(x) = v(Y)$ as, for any higher value of x, $|Q(x)| > |v(Y)|$. Also, the set $\{v.|\langle e_i, \Psi_2(A)^n \Psi_2(v) \rangle| \geqslant$

$|Q(N)|\}$ contains all reachable configurations but does not contain Y, thus (N, P) is a valid certificate.

In the general case where $\lambda \in \mathbb{C}$, we will use Property 2 to show that if there exist complex eigenvalues λ such that $|\lambda| = 1$, of multiplicity $m > 1$ with $m \neq dim(ker(A - \lambda Id))$, then $\Psi_2(A)$ admits 1 or -1 as an eigenvalue and its multiplicity $m' > 1 \neq dim(ker(\Psi_2(A) - \lambda Id))$. This implies directly the existence of at least one generalized eigenvector, thus of a family of linked left-eigenvectors of size strictly higher than 1. To this purpose, we refer to basic properties of Ψ_d:

Lemma 2.

1. $\Psi_k(A.B) = \Psi_k(A).\Psi_k(B)$
2. $\Psi_k(A^{-1}) = \Psi_k(A)^{-1}$

Proof.

1. $\Psi_k(A).\Psi_k(B)p(X) = \Psi_k(A).p(B.X) = p(A.B.X) = \Psi_k(A.B)p(X)$
2. $\Psi_k(A^{-1}).\Psi_k(A).p(X) = p(A.A^{-1}X) = p(X)$ so $\Psi_k(A^{-1}).\Psi_k(A) = Id$.

Let J the Jordan normal form of A, i.e. there exists P such that $A = P^{-1}JP$.

We have that $J = \begin{pmatrix} J_1 & 0 & \cdots & 0 \\ 0 & \ddots & \ddots & \vdots \\ \vdots & \ddots & \ddots & 0 \\ 0 & \cdots & 0 & J_k \end{pmatrix}$, and $J_k = \begin{pmatrix} \lambda_k & 1 & \cdots & 0 \\ 0 & \ddots & \ddots & \vdots \\ \vdots & \ddots & \ddots & 1 \\ 0 & \cdots & 0 & \lambda_k \end{pmatrix}$

From Lemma 2, it is easy to prove that $\Psi_d(A) = \Psi_d(P)^{-1}\Psi_d(J)\Psi_d(P)$. As $\Psi_d(A)$ and $\Psi_d(J)$ are similar, they have the same eigenvalues. We know that there exist v_1, v_2, v_3 in the base of J such that

- $v_1' = \lambda.v_1 + v_2$
- $v_2' = \lambda.v_2$
- $v_3' = \bar{\lambda}.v_3$

where v_i' is the new value of v_i in the base of J. Then the image of v_1v_3 (denoted $(v_1v_3)'$) with respect to $\Psi_2(J)$ is $v_1v_3 + \bar{\lambda}.v_2.v_3$. Also, we know that $(v_2v_3)' = v_2v_3$. Let φ such that $\varphi.\Psi_2(J).V = v_1v_3$.

$$\varphi.(\Psi_2(J) - Id)V = v_1v_3\bar{\lambda}.v_2v_3 - v_1v_3$$
$$= \bar{\lambda}.v_2v_3$$
$$\varphi.(\Psi_2(J) - Id)^2V = \bar{\lambda}.v_2v_3 - \bar{\lambda}.v_2v_3 = 0$$

As this is true for any V, then $\varphi.(\Psi_2(J) - Id) \neq 0$ and $\varphi.(\Psi_2(J) - Id)^2 = 0$. In conclusion, φ is a generalized eigenvector of $\Psi_2(J)$, thus $\Psi_2(A)$ also admits a generalized eigenvector. $\square$

Example. We consider the Orbit problem $\mathcal{O}(A, X, Y)$ with $A = \begin{pmatrix} 1 & 1 & 0 \\ 0 & 1 & 1 \\ 0 & 0 & 1 \end{pmatrix}$, $X = (-2, -1, 1)^t$ and $Y = (2, 6, 1)^t$. A admits as 1-generalized-left-eigenvectors:

$\{e_0 = (0,0,1); e_1 = (0,1,0); e_2 = (1,0,0)\}$. By the previous property, we know that $e_2 A^k = e_2 + k.e_1 + \frac{k(k-1)}{2}.e_0$, thus

$$\langle e_2 A^k, (x_X, y_X, \mathbb{1}) \rangle = y_X + k x_X + \frac{k(k-1)}{2}$$
$$= \tfrac{1}{2} k^2 - \tfrac{5}{2} k - 1$$

As we can see in Fig. 1, from $k = 3$, the value of x is strictly increasing and after $k = 7$, the value of x is strictly superior to 2. Thus we have to check a finite number of iterations before reaching $x > 2$, which is the certificate set constraint of the non-reachability of Y. For $k \in [0,6]$, Y is not reached. The couple $(7, \{(x,y,\mathbb{1}).x > 2\})$ is thus a certificate of non reachability of Y.

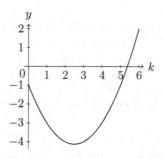

Fig. 1. Graph of the polynomial $y = \tfrac{1}{2} k^2 - \tfrac{5}{2} k - 1$

Case 4: Eigenvalues All Have a Modulus Equal to 1 and the Transformation is Diagonalizable

Some transformations do not admit generalized eigenvectors, namely diagonalizable transformations. The previous theorem is then irrelevant if for every eigenvalue λ, $|\lambda| = 1$. Such transformations are *rotations*: they remain in the same set around the origin. Take as example the transformation A of Fig. 2, taken from [8]. It defines a counterclockwise rotation around the origin by angle $\theta = \arctan(\tfrac{3}{2})$, and $\frac{\theta}{\pi}$ is not rational. The reachable set of states from X, i.e. $\{X, AX, A^2X, ...\}$ is strictly included in its closure, i.e. the set of reachable states and their neighbourhood. As Y is not on the closure of the set, then we can easily provide a non-reachability semi-algebraic invariant certificate of Y, that is the equation of the circle. However, we cannot give such a certificate for Z though it is not reachable. If it were reachable, there would exist a n such that $A^n X = Z$, thus $A^{2n} X = X$. n would also satisfy $\theta * n = 0[2\pi]$, which is impossible as $\frac{\theta}{\pi}$ is not rational. More generally, the closure of the reachable set of states of diagonalisable transformations with eigenvalues of modulus 1 is a semialgebraic set [8]. Semialgebraic certificates for such transformations exist if and only if Y does not belong to this closure [8].

Theorem 3. *For a given instance $\mathcal{O}(A, X, Y)$ such that A is diagonalizable and all its eigenvalues have a modulus of 1, eigenvectors can be used as semialgebraic certificates iff Y is not in the closure.*

$$A = \tfrac{1}{5} \begin{pmatrix} 4 & -3 \\ 3 & 4 \end{pmatrix}$$
$$X = \quad (1,0)$$
$$Y = \quad (1.5, 0.7)$$
$$Z = \quad (-1,0)$$

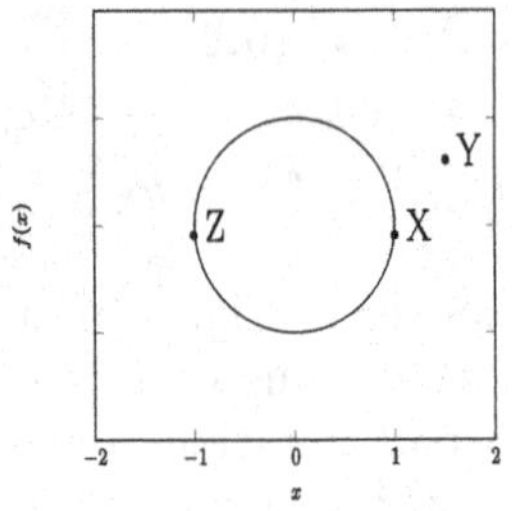

Fig. 2. Closure of the reachable set of A starting with X

Proof. Let $\mathcal{O}(A, X, Y)$ be an instance of the Orbit Problem with A a diagonaliz-able matrix only admitting eigenvalues λ such that $|\lambda| = 1$. Let φ an eigenvector of A, we denote $R = \{v | \exists k. A^k X = v\}$ the reachable set.

Lemma 3. *Let* (λ_i, φ_i) *be* d *couples of eigenvalue/left-eigenvector of a diago-nalizable matrix A of size d. Then* $R = \{v | \exists k, \forall 1 \geqslant i \geqslant d, \langle \varphi_i v, = \rangle \lambda_i^k . \langle \varphi_i, X \rangle\}$

Proof. Let $R' = \{v | \exists k, \forall 1 \geqslant i \geqslant d, \langle \varphi_i, v \rangle = \lambda_i^k \langle \varphi_i, X \rangle\}$. By the definitions of R and φ_i, the inclusion $R \subset R'$ is trivially true. Now take $v \in R'$. As there exist d different and independent eigenvectors, v is a solution of the following relation: $\exists k. \Phi v = (\lambda_1^k x_1, ... \lambda_d^k x_d)^t$, where Φ is an invertible matrix whose lines are directly defined by eigenvectors. As Φ is invertible, there exists only one solution for each k. As v is one of those solutions, then $v \in R$.

By lemma 3, for any i between 1 and d, every element v of R verifies $|\langle \varphi_i, v \rangle| = |\langle \varphi_i, X \rangle|$, thus $R \subset R_\varphi = \{v : |\langle \varphi_i, v \rangle| = |\langle \varphi_i, X \rangle|\}$. Note that this inclusion is strict, as $X' = A^{-1} X \in R_\varphi$ but $X' \notin R$. If Y does not belong to R_φ, then $(0, R_\varphi)$ is a non reachability certificate. □

3.2 General Existence of a Certificate for the Integer Orbit Problem

The Orbit Problem is originally defined on $\mathbb{Q}$. In practice, rational are not rep-resented in computers that often require the use of integers or floats. We will investigate in this section the Orbit Problem for integer transformations, i.e. matrices with coefficients in $\mathbb{Z}$. Basic matrix operations involving divisions (such as inversion) are forbidden in $\mathbb{Z}$ as it is not a field, but the only relevant oper-ation in our case is multiplication (does there exist a n such that $A^n X = Y$?) which is consistent for integer matrices.

The following property holds for integer matrices and is fundamental for the proof of the following theorem.

Property 4. *Let* $A \in mathcalM_n(\mathbb{Z})$. *If all its eigenvalue λ have a modulus inferior or equal to 1, then there exists $n > 1$ such that $\lambda^n = \lambda$.*

Proof. Let $A \in \mathcal{M}(\mathbb{Z})$ such that for all eigenvalue λ, $|\lambda| \leqslant 1$.

If $\lambda = 0$, then we can conclude right away ($0^2 = 0$).

The characteristic polynomial $P \in \mathbb{Z}[X]$ of A is monic, i.e. its leading coefficient is 1. Thus by definition, every eigenvalue is an algebraic integer. We will use the Kronecker theorem [15], stating that if a non null algebraic integer α has all its rational conjugates (i.e. roots of its rational minimal polynomial) admitting a modulus lower or equal to 1, then α is a root of unity.

Each eigenvalue λ admits a minimal rational polynomial Q. We can show that Q necessarily divides P by performing an euclidian division : there exist $D, R \in \mathbb{Q}[X]$ such that $P(X) = Q(X)D(X)+R(X)$, with the degree of R strictly inferior to Q. We know that $P(\lambda) = 0$ and $Q(\lambda) = 0$, thus $R(\lambda) = 0$. If $R \neq 0$, then R is the minimal polynomial of λ as its degree is inferior to the degree of Q, which is absurd by hypothesis. Thus, the set of rational conjuguates of λ are roots of P, by hypothesis of modulus inferior or equal to 1. By the Kronecker theorem, λ is a root of unity, i.e. $\exists n > 1.\lambda^n = \lambda$. $\qquad\square$

Theorem 4. *Any non-reachable instance of the Orbit problem $\mathcal{O}(A, X, Y)$ where $A \in \mathcal{M}_n(\mathbb{Z})$ admits a closed semi-algebraic certificate.*

Because of length constraints, the proof of this theorem cannot fit in this article. It can be found on an extended version of this paper [6]. The idea is to consider that if all eigenvalues have a modulus of 1, then by Property 4, the reachable set of states is finite in any case.

4 Conclusion and Future Work

This paper presents new insights on the quality of certificates necessary to prove the non-reachability of a given Orbit problem instance. In addition, in contrast with [8], we gain simplicity and precision by not studying the Jordan normal form of a linear transformation but only its eigenvector decomposition.

Eigenvectors are computable without knowledge of the initial state X and the target Y. It means that certificates are intrinsequely linked only to the transformation studied. In other words, for an instance of the Orbit Problem $\mathcal{O}(A, X, Y)$, X and Y play a minor role in the expression of certificates. As a consequence, generalizing the result of this paper to sets of initial states and targets should be possible.

As this article explores the Orbit Problem for rationals, it is worth noting that certificates may not necessarily be relevant for real-life programs manipulating floats. For example, the Orbit problem $(x \mapsto \frac{x}{2}, 1, 0)$ has a solution for some floating point implementations due to limited precision. The question of certificates synthesis for such problems is also an interesting challenge.

References

1. Blazy, S., Bühler, D., Yakobowski, B.: Structuring abstract interpreters through state and value abstractions. In: Bouajjani, A.; Monniaux, D. (eds.) VMCAI 2017. LNCS, vol. 10145, pp. 112–130. Springer, Cham (2017). https://doi.org/10.1007/978-3-319-52234-0_7

2. Bozga, M., Iosif, R., Konečný, F.: Fast acceleration of ultimately periodic relations. In: Touili, T., Cook, B., Jackson, P. (eds.) CAV 2010. LNCS, vol. 6174, pp. 227–242. Springer, Heidelberg (2010). https://doi.org/10.1007/978-3-642-14295-6_23

3. Cousot, P., Cousot, R.: Abstract interpretation: a unified lattice model for static analysis of programs by construction or approximation of fixpoints. In: Proceedings of the 4th ACM SIGACT-SIGPLAN Symposium on Principles of Programming Languages, pp. 238–252. ACM (1977)

4. de Oliveira, S., Bensalem, S., Prevosto, V.: Polynomial invariants by linear algebra. In: Artho, C., Legay, A., Peled, D. (eds.) ATVA 2016. LNCS, vol. 9938, pp. 479–494. Springer, Cham (2016). https://doi.org/10.1007/978-3-319-46520-3_30

5. de Oliveira, S., Bensalem, S., Prevosto, V.: Synthesizing invariants by solving solvable loops. In: D'Souza, D., Narayan Kumar, K. (eds.) ATVA 2017. LNCS, vol. 10482, pp. 327–343. Springer, Cham (2017). https://doi.org/10.1007/978-3-319-68167-2_22

6. de Oliveira, S., Prevosto, V., Habermehl, P., Bensalem, S.: Left-eigenvectors are certificates of the orbit problem. http://steven-de-oliveira.fr/content/publis/certificates_2018.pdf

7. Ernst, M.D., Cockrell, J., Griswold, W.G., Notkin, D.: Dynamically discovering likely program invariants to support program evolution. IEEE Trans. Softw. Eng. **27**(2), 99–123 (2001)

8. Fijalkow, N., Ohlmann, P., Ouaknine, J., Pouly, A., Worrell, J.: Semialgebraic invariant synthesis for the Kannan-Lipton orbit problem. In: STACS 2017. LIPIcs, vol. 66, pp. 29:1–29:13. Schloss Dagstuhl - Leibniz-Zentrum fuer Informatik (2017)

9. Filliâtre, J.-C., Paskevich, A.: Why3—where programs meet provers. In: Felleisen, M., Gardner, P. (eds.) ESOP 2013. LNCS, vol. 7792, pp. 125–128. Springer, Heidelberg (2013). https://doi.org/10.1007/978-3-642-37036-6_8

10. Kannan, R., Lipton, R.J.: The orbit problem is decidable. In: Proceedings of the Twelfth Annual ACM Symposium on Theory of Computing, pp. 252–261. ACM (1980)

11. Kannan, R., Lipton, R.J.: Polynomial-time algorithm for the orbit problem. J. ACM (JACM) **33**(4), 808–821 (1986)

12. Kovács, L.: Reasoning algebraically about P-solvable loops. In: Ramakrishnan, C.R., Rehof, J. (eds.) TACAS 2008. LNCS, vol. 4963, pp. 249–264. Springer, Heidelberg (2008). https://doi.org/10.1007/978-3-540-78800-3_18

13. Rocha, H., Ismail, H., Cordeiro, L., Barreto, R.: Model checking embedded C software using k-induction and invariants. In: 2015 Brazilian Symposium on Computing Systems Engineering (SBESC), pp. 90–95. IEEE (2015)

14. Rodríguez-Carbonell, E., Kapur, D.: Generating all polynomial invariants in simple loops. J. Symb. Comput. **42**(4), 443–476 (2007)

15. Schinzel, A., Zassenhaus, H.: A refinement of two theorems of Kronecker. Michigan Math. J **12**, 81–85 (1965)

Constrained Dynamic Tree Networks

Matthew Hague[1][(✉)] and Vincent Penelle[2]

[1] Royal Holloway, University of London, Egham, UK
matthew.hague@rhul.ac.uk
[2] Université de Bordeaux, LaBRI, UMR 5800, Talence, France
vincent.penelle@labri.fr

Abstract. We generalise Constrained Dynamic Pushdown Networks, introduced by Bouajjani *et al.*, to Constrained Dynamic Tree Networks. In this model, we have trees of processes which may monitor their children. We allow the processes to be defined by any computation model for which the alternating reachability problem is decidable. We address the problem of symbolic reachability analysis for this model. More precisely, we consider the problem of computing an effective representation of their reachability sets using finite state automata. We show that backwards reachability sets starting from regular sets of configurations are always regular. We provide an algorithm for computing backwards reachability sets using tree automata.

Keywords: Model-checking · Dynamic networks · Concurrency
Pushdown systems · Alternation · Higher-order
Collapsible pushdown systems

1 Introduction

Bouajjani *et al.* [2] defined Constrained Dynamic Networks of Pushdown Systems: a model of concurrent computation where configurations of processes are tree structures, and each process is given by a pushdown system. During an execution, new child processes can be created, and a parent can test the states of its children before performing an execution step. They considered the global backwards reachability problem for these systems. That is, given a regular set of target configurations, compute the set of configurations that can reach the target set. They showed that, under a *stability* constraint, this backwards reachability set is regular and computable.

The stability constraint requires that once a test a parent may make on its children is satisfied, then it will remain satisfied, even if the children continue their execution. In the simplest case, this allows a parent to test for termination in a given state of its children. In general, this constraint allows a parent to (repeatedly) test whether its children have passed certain stages of execution (and their state in doing so).

We show here that Bouajjani *et al.*'s result is not dependent on the processes in the tree being modelled by pushdown systems. In fact, all that is required is

I. Potapov and P.-A. Reynier (Eds.): RP 2018, LNCS 11123, pp. 45–58, 2018.
https://doi.org/10.1007/978-3-030-00250-3_4

that the *alternating reachability problem* is decidable for the systems labelling the nodes in the tree. Intuitively, in the alternating reachability problem, some steps during the run of the system may be required to split into separate paths. From the initial state, we ask whether all paths of the execution reach a given final state.

Thus, we introduce *Constrained Dynamic Tree Networks*, which are tree networks of processes as before, but the individual processes can be labelled by any system for which the alternating reachability problem is decidable.

One particular instance of interest is the case of networks of *collapsible pushdown systems* [14]. Collapsible pushdown systems are a generalisation of pushdown systems that are known to be equi-expressive with *Higher-Order Recursion Schemes*. The alternating reachability problem is known to be decidable for these systems [32]. In fact, the backwards reachability sets of alternating collapsible pushdown systems are also known to be computable and regular [4]. Thus, we obtain a new model of concurrent higher-order programs for which the backwards reachability sets are also computable and regular. An advantage of our approach is that we do not need to consider the technical difficulties of reasoning about collapsible pushdown systems. The proof presented here only needs to take care of the concurrent aspects of the computations. Thus, we obtain results for quite complex systems with a relatively modest proof.

Modern day programming increasingly embraces higher-order programming, both via the inclusion of higher-order constructs in languages such as C++, JavaScript and Python, but also via the importance of *callbacks* in highly popular technologies such as jQuery and Node.js. For example, to read a file in Node.js, one would write

```
fs.readFile('f.txt', function (err, data) { ..use data.. });
```

In this code, the call to `readFile` spawns a new thread that asynchronously reads `f.txt` and sends the `data` to the function argument. This function will have access to, and frequently use, the closure information of the scope in which it appears (for example, variables defined before the `readFile` statement). The rest of the program runs *in parallel* with this call. This style of programming is fundamental to both jQuery and Node.js programming, as well as being a popular for programs handling input events or slow IO operations such as fetching remote data or querying databases (e.g. HTML5's indexedDB).

Analysing such programs is a challenge for verification tools which usually do not model higher-order recursion, or closures, accurately. However, several higher-order model-checking tools have been recently developed. This trend was pioneered by Kobayashi *et al.* [18]. The feasibility of higher-order model-checking in practice has been demonstrated by numerous higher-order model-checking tools [5,6,17,19,21,30,36]. Since all of these tools can handle the alternating reachability problem, it is possible that our techniques may be used to provide model checking tools for concurrent higher-order programs.

Our construction follows Bouajjani *et al.* and uses a *saturation* method to construct a regular representation of the backwards reachability set. However, our automaton representation is different: it separates the representation of the system states from the tree structure. We also use different techniques to prove correctness of the construction. In particular, our soundness proof works by defining and showing soundness of each transition of the automaton, rather than dissecting complete runs. This is an application of a technique first used for a saturation technique for solving parity games over pushdown systems [15].

The full version of this article with appendices is available online [16].

1.1 Related Work

Dynamic pushdown networks have been studied without the process tree structure or constraints allowing a parent to inspect its children [26,39]. Various decidability-preserving locking techniques have also been investigated [24]. Some of these works also allow the tree structure to be taken into account [23,31]. Touili and Atig have also considered communication structures that are not necessarily trees [41]. However, these works consider pushdown networks only.

There has been some work studying concurrent variants of recursion scheme model checking, including a context-bounded algorithm for recursion schemes [20], and further underapproximation methods such as phase-bounded, ordered, and scope-bounding [12,38]. These works allow only a fixed number of threads.

Dynamic thread creation is permitted by both Yasukata et al. [42,43] and by Chadha and Viswanathan [7]. In Yasukata et al.'s model, recursion schemes may spawn and join threads. Communication is permitted only via nested locks. Their work is a generalisation of results for order-1 pushdown systems [11]. Chadha and Viswanathan allow threads to be spawned, but only one thread runs at a time, and must run to completion. Moreover, the tree structure is not maintained.

The saturation technique was popularised by Bouajjani et al. [1] for the analysis of pushdown systems, which was implemented in the successful Moped tool [37,40]. Saturation methods also exist for *ground tree rewrite systems* and related systems [3,25,27], though use different techniques.

Ground tree rewrite systems may also be generalised to trees where the nodes are labelled by higher-order stacks. Penelle proves decidability of first order logic with reachability over such systems [34]. However, this result does not allow nodes to have an unbounded number of direct children, and does not consider collapsible stacks in their full generality.

A related model of tree rewriting was introduced by Clemente et al. [8]. This model allows more powerful rewriting rules than ground tree rewrite systems while still enjoying decidability of alternating reachability. It is shown that reachability over alternating variants of a number of pushdown system models can be reduced to this model. In particular, this includes (ordered) annotated pushdown systems which are tightly related to (concurrent) collapsible pushdown systems. We believe it is likely that constrained dynamic networks of annotated pushdown systems could also be encoded in this model. However, our result applies to any system for which alternating reachability is decidable, and does not require an encoding of the underlying model into any particular form.

There is various research into meta-results on the analysis of concurrent systems, where the concurrent structure is the object of the research. Recent work by La Torre *et al.* has shown that parameterised safety analysis is possible of asynchronous networks of shared-memory systems [22], provided, amongst other constraints, the downwards closure of the system is computable. Muscholl *et al.* also consider a parameterised model where processes may spawn an arbitrary (uncontrolled) amount of identical child processes [29]. Collapsible pushdown systems are known to have these properties [9,13,33,44]. Other works have studied multi-stack pushdown systems and offered either bounded tree-width [28], or split-width [10], as explanations for decidability. However, these results have not been extended to higher orders.

2   Alternating Transition System

We define the notion of alternating transition system. An alternating transition system accepts labels γ to which operations θ can be applied. Transitions of the system are of the form $s \xrightarrow{\theta} S$. Where S is a set of states. A label γ is accepted from s whenever $\theta(\gamma)$ is accepted from every state in S. If a state s is final, it accepts all labels.

We will consider Γ to be a set of labels, and Ops a set of operations $\theta : \Gamma \nrightarrow \Gamma$ over Γ. Note, we do not require that θ is defined over all elements of Γ. We also define the special operation Id such that $\mathrm{Id}(\gamma) = \gamma$ for all $\gamma \in \Gamma$.

Definition 1 (Alternating transition systems over Γ, Ops). *An alternating transition system over Γ, Ops is a tuple $\mathcal{N} = (\mathbb{S}, \mathbb{F}, \eta)$, where $\mathbb{S}$ is a finite set of states, $\mathbb{F} \subseteq \mathbb{S}$ is the set of final states, $\eta \subseteq \mathbb{S} \times \mathrm{Ops} \times 2^{\mathbb{S}}$ is the set of transitions.*

Given $\gamma \in \Gamma$ and $s \in \mathbb{S}$, we inductively define acceptance of γ from s, denoted $\gamma \vdash s$. We have $\gamma \vdash s$ if s is final, or if there is a transition $\nu = (s, \theta, S)$ such that $\theta(\gamma)$ is defined and $\theta(\gamma) \vdash s'$ for every $s' \in S$.

Requirement. In all the following, we suppose that for every alternating transition system $\mathcal{N}$, given $\gamma \in \Gamma$ and $s \in \mathbb{S}$, we can decide whether $\gamma \vdash s$.

Example 1. An alternating pushdown system with stack alphabet Σ is an alternating transition system with $\Gamma = \Sigma^*$ and the set of operations Ops $= \{(a, u) \mid a \in \Sigma, u \in \Sigma^*\}$. Where for all $w \in \Sigma^*$

$$(a, u)(w) = \begin{cases} uv & w = av \\ \text{undefined} & \text{otherwise.} \end{cases}$$

Here we represent a stack as a word, and the top of the stack appears leftmost. A transition $s \xrightarrow{(a,u)} S$ represents an alternating transition from a configuration (s, aw) of the pushdown system (with control state s and stack aw) to a set of configurations containing (s', uw) for each $s' \in S$. We will have $aw \vdash s$ if we can show $s' \vdash uw$ for each $s' \in S$.

3   Constrained Dynamic Tree Networks

We define constrained dynamic tree networks (CDTNs), which allow process trees with dynamic thread creation and parents to inspect their children.

Definition 2 (Constrained Dynamic Tree Network over Γ, Ops). *A constrained dynamic network over Γ, Ops is a tuple $\mathcal{M} = (\mathrm{P}, \mathrm{F}, \delta)$ with:*

- *P is a finite set of states and $\mathrm{F} \subseteq \mathrm{P}$ is the set of final states,*
- *δ a finite set of transitions of the following form:*
  C1 *$\phi : p \xrightarrow{\theta} p_a$, with $\theta \in \mathrm{Ops}, p, p_a \in \mathrm{P}$, and ϕ is a regular language over P^*.*
  C2 *$\phi : p \to p_a \rhd p_b$, with $p, p_a, p_b \in \mathrm{P}$, and ϕ is a regular language over P^*.*

An $\mathcal{M}$-configuration is a tree labelled by $\mathrm{P} \times \Gamma$. Let $\mathcal{T}(\mathrm{P} \times \Gamma)$ denote the set of these configurations. More explicitly, a configuration is either a leaf node $(p, \gamma)(\emptyset)$ or a tree $(p, \gamma)(t_1, \cdots, t_m)$ with root (p, γ) and children $t_1, \cdots, t_m$ where $p \in \mathrm{P}$, $\gamma \in \Gamma$, and for each $1 \le i \le m$ we have that t_i is an $\mathcal{M}$-configuration. A *context* C is a tree labelled by $(\mathrm{P} \times \Gamma) \cup \{\square\}$ containing exactly one node labelled by $\square$, which is a leaf. We write $C[t]$ to denote the configuration obtained by replacing $\square$ by t in C. Furthermore, let

$$S((p, \gamma)(t_1, \cdots, t_m)) = p$$

extract the *internal* state of the root node of a configuration.

Transitions of the form C1 apply θ to a node, while transitions of the form C2 create a new child process. That is, the application of a transition of the form C1 to a configuration $C[(p, \gamma)(t_1, \cdots, t_m)]$ yields $C[(p_a, \theta(\gamma))(t_1, \cdots, t_m)]$, if $\theta(\gamma)$ is defined and $S(t_1) \cdots S(t_m) \in \phi$. The application of a transition of the form C2 to a configuration $C[(p, \gamma)(t_1, \cdots, t_m)]$ yields the configuration $C[(p_a, \gamma)(t_1, \cdots, t_m, (p_b, \gamma)(\emptyset))]$ if $S(t_1) \cdots S(t_m) \in \phi$.

3.1   Stability Constraint

We give the restriction on child constraints ϕ that allows us to preserve decidability of reachability for CDTNs. Intuitively, this constraint asserts that once a constraint ϕ is satisfied, it will remain satisfied even if its children progress.

Definition 3 (Stability relation [2]). *Given an alphabet Σ and a binary relation ρ over Σ, we say that a subset $\mathfrak{S}$ of Σ is ρ-stable if for every $a, b \in \Sigma$, $\rho(a, b) \wedge a \in \mathfrak{S} \Rightarrow b \in \mathfrak{S}$.*

A language L is ρ-stable if it is defined by a regular expression of the form

$$e ::= \mathfrak{S}, \rho\text{-stable set} \mid e + e \mid e.e \mid e^*$$

In [2], it is shown that if a language L is ρ-stable, for every $a, b \in \Sigma, u, v \in \Sigma^*$, $uav \in L \wedge \rho(a, b) \Rightarrow ubv \in L$. Given a CDTN $\mathcal{M} = (\mathrm{P}, \mathrm{F}, \delta)$ we define

$$\rho_\delta = \{(p, p') \mid \exists \phi : p \xrightarrow{\theta} p' \in \delta \vee \exists \phi : p \to p' \rhd p'' \in \delta\} .$$

We say $\mathcal{M}$ is ρ_δ-stable iff for all $\phi : p \xrightarrow{\theta} p_a \in \delta$ and $\phi : p \to p_a \rhd p_b \in \delta$ we have ϕ is ρ_δ-stable (can be checked looking at regular expressions defining each ϕ).

3.2   Automaton

We now define a notion of tree automata over the configurations of a constrained dynamic tree network. As these configurations can have an unbounded arity, we need to have an automaton model which can deal with unbounded arity, thus we use an adapted version of hedge automata. Transitions of our automata are of the form $p(L) \to q$, meaning they can rewrite a tree to a state q, if

- the internal state of its root is p,
- the ith son of its root can be rewritten to the state q_i, and
- $q_1 \cdots q_m$ is in the regular language L (if the node has m sons).

Moreover, the automaton checks that the element of the root is accepted by an alternating transition system which is bound to the transition (more precisely, we will use a single alternating transition system for the whole automaton, which has a unique initial state for each rule of the automaton). In the following definition, let $\mathrm{Reg}(\mathcal{Q})$ be the set of regular languages over alphabet $\mathcal{Q}$.

Definition 4 ($\mathcal{M}$-automaton). *Given a CDTN $\mathcal{M} = (\mathrm{P}, \mathrm{F}, \delta)$, an $\mathcal{M}$-automaton is a tuple $\mathcal{A} = (\mathcal{Q}, \mathcal{F}, \Delta, \mathcal{N})$, where:*

- *$\mathcal{Q}$ is a finite set of states and $\mathcal{F} \subseteq \mathcal{Q}$ the set of final states,*
- *$\Delta \subseteq \mathrm{P} \times \mathrm{Reg}(\mathcal{Q}) \times \mathcal{Q}$ a finite set of transitions of the form $p(L) \to q$,*
- *$\mathcal{N} = (\mathbb{S}, \mathbb{F}, \eta)$ an alternating transition system over Γ, Ops, such that for every $r \in \Delta$, there is a unique state $s_r \in \mathbb{S}$. Without loss of generality, we suppose that these states have no incoming transition[1] and that these states are not final[2]. Intuitively, s_r accepts the set of elements of Γ that allow r to fire. A $\mathcal{M}$-automaton is analogous to a tree automaton, with the difference that letters are replaced with sets of labels accepted from a state of an alternating transition system.*

An $\mathcal{A}$-configuration is a tree labelled by $(\mathrm{P} \times \Gamma) \cup \mathcal{Q}$, such that only leaves can be labelled by $\mathcal{Q}$. Given a transition $r = p(L) \to q$ and two $\mathcal{A}$-configurations t and t', we have $t \xrightarrow{r} t'$ if and only if $t = C[(p, \gamma)(q_1, \cdots, q_m)]$, $t' = C[q]$, $q_1 \cdots q_m \in L$ and $\gamma \vdash s_r$.

Let $\xrightarrow[\Delta]{*}$ be the transitive closure of $\left(\bigcup_{r \in \Delta} \xrightarrow{r} \right)$. The set of $\mathcal{M}$-configurations recognised by $\mathcal{A}$ from the state q is $\mathcal{L}_q(\mathcal{A}) = \{ t \in \mathcal{T}(\mathrm{P} \times \Gamma) \mid t \xrightarrow[\Delta]{*} q \}$.

Note, the membership problem for $\mathcal{M}$-automata is decidable whenever it is decidable whether $\gamma \vdash s$ for a given γ and s. Similarly, emptiness is decidable whenever it is decidable if $\exists \gamma . \gamma \vdash s$ for a given s.

---

[1] If it is not the case, we create a copy of these states on which we conserve all the transition as an "internal state", and remove the incoming transitions to these states.

[2] If so, for a state s_r, we create a new final state s and add the transition $s_r \xrightarrow{\mathrm{Id}} s$, and remove s_r from the set of final states.

Example 2. We can accept regular sets of pushdown networks as defined by Bouajjani *et al.* [2] by defining the word automata used to recognise pushdown stacks as alternating transition systems with operations of the form (a, ε), where ε is the empty word, and operations have the same semantics as in Example 1. That is, each operation consumes the leftmost character of the word representation of the stack. For this we will need an explicit end-of-stack marker.

4   Backwards Reachability

In this section, we show that we can compute the backwards reachability set of CDTNs. That is, if a CDTN $\mathcal{M}$ is ρ_δ-stable, then the set of predecessors of a regular set is regular. Here, by regular we mean the set is accepted by an $\mathcal{M}$-automaton. We remark in the conclusion how this notion of regularity may be related to a more conventional one.

Given S a set of $\mathcal{M}$-configurations, we denote $\mathrm{pre}^*_{\mathcal{M}}(S)$ the set of *predecessors* of elements of S, i.e., $\mathrm{pre}^*_{\mathcal{M}}(S) = \{s \mid \exists s' \in S, s \xrightarrow[\mathcal{M}]{*} s'\}$.

Theorem 1. *Given $\mathcal{M}$ a ρ_δ-stable CDTN and $\mathcal{A}$ an $\mathcal{M}$-automaton, it is possible to compute a $\mathcal{M}$-automaton $\mathcal{A}'$ such that $\mathcal{L}(\mathcal{A}') = \mathrm{pre}^*_{\mathcal{M}}(\mathcal{L}(\mathcal{A}))$.*

For the proof, we construct the automaton $\mathcal{A}'$ from $\mathcal{A}$ and $\mathcal{M}$ in two steps.

4.1   The Automaton $\mathcal{A}_p$

First we add to the states of the automaton the internal state of the root of the $\mathcal{M}$-configuration that was reduced to this state. Informally, we replace every transition $p(L) \to q$ with $p(L) \to (q, p)$, so given an $\mathcal{M}$-configuration t such that if $t \xrightarrow[\Delta]{*} q$, we have $t \xrightarrow[\Delta_p]{*} (q, S(t))$. This will be useful in the actual construction of $\mathcal{A}'$, as to inversely apply $\mathcal{M}$-rules, we will need to check if the constraint of the rule is satisfied, which will be given by this information (using the stability property, as we remember the final state of the root of each son). More formally, we will also need to adapt the constraint L and to add states to the inner alternating transition system. For notational convenience, let $\mathcal{Q}_p = Q \times \mathrm{P}$.

We define $\mathcal{A}_p = (\mathcal{Q}_p, \mathcal{F} \times \mathrm{P}, \Delta_p, \mathcal{N}_p)$, where

$$\Delta_p = \left\{ p(L_\mathrm{P}) \to (q, p) \;\middle|\;
\begin{array}{l}
p(L) \to q \in \Delta, \\
L_\mathrm{P} = \left\{ (q_1, p_1) \cdots (q_m, p_m) \;\middle|\; \begin{array}{l} q_1 \cdots q_m \in L, \\ p_1, \cdots, p_m \in \mathrm{P} \end{array} \right\}
\end{array}
\right\}$$

and $\mathcal{N}_p = (\mathbb{S}_p, \mathbb{F}_p, \eta_p)$, with

- $\mathbb{S}_p = \mathbb{S} \backslash \{s_r \mid r \in \Delta\} \cup \{s_r \mid r \in \Delta_p\}$,
- $\mathbb{F}_p = \mathbb{F} \cap \mathbb{S}_p \cup \{s_r \mid r = p(L_\mathrm{P}) \to (q, p), s_{r'} \in \mathbb{F}, r' = p(L) \to q\}$,
- $\eta_p = \eta \cup \{s_r \xrightarrow{\theta} S \mid s_{r'} \xrightarrow{\theta} S \in \eta, r = p(L_\mathrm{P}) \to (q, p), r' = p(L) \to q\}$.

Lemma 1. $\mathcal{L}(\mathcal{A}_p) = \mathcal{L}(\mathcal{A})$.

Proof. We only have to observe that for every t, $t \xrightarrow[\Delta_p]{*} (q, S(t))$ if and only if $t \xrightarrow[\Delta]{*} q$, and that (q, p) is final if and only if q is final.

4.2   From Constraints over P to Constraints over $\mathcal{Q}_p$

In order to faithfully compute the automaton $\mathcal{A}'$, we need to be able to transfer the constraint of $\mathcal{M}$ to the states of $\mathcal{A}'$. Indeed, we need to recognise only valid predecessors of the configurations recognised by $\mathcal{A}$, i.e. those which satisfy the constraints ϕ. Given a regular language $\phi \subseteq P^*$, we thus define $\langle \phi \rangle = \{(q_1, p_1) \cdots (q_m, p_m) \mid p_1 \cdots p_m \in \phi, q_1, \cdots, q_m \in \mathcal{Q}\}$. It is straightforward to see that this language is also regular.

4.3   Closed Set of Constraints

During the construction of $\mathcal{A}'$, we add new transitions of the form $p(L) \to (q', p')$. The constraints L will be constructed from those already appearing in $\mathcal{A}_p$ and the constraints ϕ used in $\mathcal{M}$, using intersection and right-quotient operations. Intersection $L \cap \langle \phi \rangle$ allows us to check that the guarding constraint of an $\mathcal{M}$-rule is satisfied at the considered position in the configuration. The right-quotient

$$L(q, p)^{-1} = \{(q_1, p_1) \cdots (q_m, p_m) \mid (q_1, p_1) \cdots (q_m, p_m)(q, p) \in L\}$$

allows us to get immediate predecessors by an operation of the form C2. We define Λ to be the smallest family of languages over $\mathcal{Q}_p$ such that:

- If $r = p(L) \to (q, p) \in \Delta_p$, then $L \in \Lambda$,
- If $L \in \Lambda$ and $\tau = \phi : p \xrightarrow{\theta} p_a \in \delta$, or $\tau = \phi : p \to p_a \triangleright p_b \in \delta$, then $L \cap \langle \phi \rangle \in \Lambda$,
- If $L \in \Lambda$ and $(q, p) \in \mathcal{Q}_p$, then $L(q, p)^{-1} \in \Lambda$.

Finiteness of Λ was shown by Bouajjani *et al.* [2]. To prove it, observe that as the L and ϕ are regular, there are automata recognising them. Moreover there is a finite number of such constraints. We can take the product of all these automata to get a finite automaton, and associate each constraint with a set of final states of the product. Indeed, each $L \in \Lambda$ can be associated with a set of final states (as taking the right-product is equivalent to moving backward by one transition, and as we already have a product automaton, we don't have to introduce new states for the intersection). Thus, only a finite number of automata can be generated.

Lemma 2. [2, Lemma 3] Λ *is finite.*

4.4   Constructing $\mathcal{A}'$

We now actually describe our saturation algorithm constructing $\mathcal{A}'$. To do so we start from $\mathcal{A}_p$ and only add new transitions: we will never add new states, so this process terminates. The main idea is, for every $\mathcal{M}$-rule $r = \phi : p \xrightarrow{\theta} p_a$ and every transition $p_a(L) \to (q', p')$ starting with p_a, to add a new transition starting with p and ending in the same states (q', p'). Moreover, we ensure the sons of the node we apply the rule to satisfy ϕ by setting the constraint of the rule to $L \cap \langle \phi \rangle$. We also ensure that the elements recognised from the state associated with the new rule are predecessors by θ of those recognised from the one associated with the old rule. For the spawning rule $\phi : p \to p_a \rhd p_b$, we moreover ensure that there is exactly one son less and the label was also accepted by the last son. We need that the label was also accepted by the last son since the spawn operation creates a copy of the parent process's label. Hence, the label of the parent must also be the label of the last son.

We construct $\mathcal{A}' = (\mathcal{Q} \times \mathrm{P}, \mathcal{F} \times \mathrm{P}, \Delta', \mathcal{N}')$, with $\mathcal{N}' = (\mathbb{S}_p, \mathbb{F}_p, \eta')$. We give the formal definition of the construction first, and then informally explain the two rules R1 and R2.

We define Δ' and η' inductively as the fixed point of the following sequence. We begin with $\Delta'_0 = \Delta_p$ and $\eta'_0 = \eta_p$. Now, suppose Δ'_{i-1} and η'_{i-1} are defined. We construct Δ'_i and η'_i to be at least Δ'_{i-1} and η'_{i-1} plus transitions added with one of the following rules:

R1. if we have:
- $\tau = \phi : p \xrightarrow{\theta} p_a \in \delta$,
- $r = p_a(L) \to (q', p') \in \Delta'_{i-1}$,

we add
- $r' = p(L \cap \langle \phi \rangle) \to (q', p')$ to Δ'_i,
- $\nu' = s_{r'} \xrightarrow{\theta} \{s_r\}$ to η'_i.

R2. if we have:
- $\tau = \phi : p \to p_a \rhd p_b \in \delta$,
- $r_1 = p_a(L_1) \to (q', p') \in \Delta'_{i-1}$,
- $r_2 = p_b(L_2) \to (q'', p'') \in \Delta'_{i-1}$, with $\varepsilon \in L_2$,

we add
- $r' = p(L_1(q'', p'')^{-1} \cap \langle \phi \rangle) \to (q', p')$ to Δ'_i,
- $\nu' = s_{r'} \xrightarrow{\mathrm{Id}} \{s_{r_1}, s_{r_2}\}$ to η'_i.

This process terminates when $\Delta'_{i-1} = \Delta'_i$ and $\eta'_{i-1} = \eta'_i$. As the set of states is fixed, there is a finite number of possible rules, thus we terminate and $\mathcal{A}'$ exists.

Intuitively, R1 works as follows. We want to extend the automaton to recognise the result of a reverse application of $\phi : p \xrightarrow{\theta} p_a$. That is, whenever a configuration t' with the root node having internal state p_a is accepted, we should now accept a configuration t with root internal state p. Hence, we look for a transition (r) that will read and accept the root node of t' and introduce a new transition (r') that will read and accept the root of t. In addition, we need to take care of the children of the root. In particular, to be able to apply τ the children must satisfy ϕ. This is why we intersect with $\langle \phi \rangle$. Furthermore, to simulate the (reverse) update to γ, we add the transition $s_{r'} \xrightarrow{\theta} \{s_r\}$ to assert that the label accepted by r' would be accepted by r after an application of θ.

The rule R2 works similarly to R1, except we need to deal with the addition of a new child in the transition from t to t'. This is a removal when applied in reverse, hence the introduced transition performs a right-quotient on the language of children. In addition, we have to ensure that the spawned child has the same label as the parent. To do this, we look at the transition r_2 used to accept the final child. Note, the right quotient removes the target (q'', p'') of this transition. When applying this transition is reverse, the label γ of the root of t must be the same as the label of the root of t' and its final child. This explains the transition $s_{r'} \xrightarrow{\text{Id}} \{s_{r_1}, s_{r_2}\}$ which ensures γ is accepted at both the root and its final child.

5   Correctness

We show that $\mathcal{A}'$ accepts $\text{pre}^*_{\mathcal{M}}(\mathcal{L}(\mathcal{A}))$. It is sufficient to prove the following property, which we discuss in the following subsections.

Proposition 1. *Given* $(q, p) \in \mathcal{Q}_p$, *we have* $\mathcal{L}_{(q,p)}(\mathcal{A}') = \text{pre}^*_{\mathcal{M}}(\mathcal{L}_{(q,p)}(\mathcal{A}_p))$.

5.1   Soundness

Proposition 2. *Given* $(q, p) \in \mathcal{Q}_p$, *we have* $\mathcal{L}_{(q,p)}(\mathcal{A}') \subseteq \text{pre}^*_{\mathcal{M}}(\mathcal{L}_{(q,p)}(\mathcal{A}_p))$.

We give the complete proof of this proposition in the full version. Intuitively, to prove this proposition, we associate to each state of an automaton (and the inner alternating transition system as well) a *meaning* that is intimately connected to the backwards reachability set we want to construct. We consider a transition $r = p(L) \rightarrow (q, p')$ to be sound under the following condition: if we take elements satisfying the meaning of each state appearing at the left of the transition, then the configuration including all these elements satisfies the meaning of the right state of the transition. Intuitively this says that, assuming all actions taken by other transitions in the automaton are correct, the current transition does nothing wrong. We inductively show that every transition appearing in $\mathcal{A}'$ is sound. Finally, we show that if an automaton is sound and contains $\mathcal{A}_p$, it satisfies the proposition, showing that it is the case for $\mathcal{A}'$.

5.2   Completeness

The proof of completeness of $\mathcal{A}'$ is conceptually simpler than the soundness proof. It proceeds by a straightforward induction over the length of the run showing a configuration is in the backwards reachability set. In the base case we have the configuration is accepted by $\mathcal{A}_p$ and the proof is immediate. In the inductive case, we have t reaches t' by a single transition, and an accepting run of $\mathcal{A}'$ over t'. We then inspect the transition from t to t' and show that our construction of $\mathcal{A}'$ ensures that we can modify the accepting run of t' to obtain an accepting run of t. For space reasons, we give the proof in the full version.

Proposition 3. *Given* $(q, p) \in \mathcal{Q}_p$, *we have* $\text{pre}^*_{\mathcal{M}}(\mathcal{L}_{(q,p)}(\mathcal{A}_p)) \subseteq \mathcal{L}_{(q,p)}(\mathcal{A}')$.

6   Conclusion

We have shown that the saturation algorithm for constrained dynamic pushdown networks introduced by Bouajjani *et al.* [2] can be generalised to not only pushdown networks, but networks of any system for which the alternating reachability problem is decidable. In particular, this includes collapsible pushdown systems, or higher-order recursion schemes, which thus allows the analysis of a kind of concurrent higher-order programs.

We showed that, given a target set of configurations represented by an $\mathcal{M}$-automata, the backwards reachability set is computable and also representable by an $\mathcal{M}$-automaton. We make some remarks on $\mathcal{M}$-automata as a notion of regularity. In order to accept a configuration, an automaton must perform several alternating reachability checks. This is not regular in the conventional sense. However, for alternating pushdown systems, and indeed alternating collapsible pushdown systems, the backwards reachability set of a regular set of stacks is known to have a regular representation [1,4]. Thus, we can replace the alternating reachability tests with regular automata which run over the stack contents labelling each node. Thus we obtain a truly regular representation of the backwards reachability sets of CDTNs over these systems.

A natural avenue of future work is to attempt to generalise our model further, to permit more intricate communication between processes. One option is to allow the child nodes to inspect the internal state of their parent processes. In general this leads to an undecidable model. It is an open problem to discover a form of interesting upwards communication that is decidable. Similarly, we may seek to relax the stability constraint. One such option is to use the stability constraint defined by Touili and Atig [41] where internal states are grouped into mutually reachable equivalence classes. Thus, any run moves through a bounded number of equivalence classes. We can then insist that constraints are over the equivalence classes rather than individual states. This is reminiscent of *context-bounded* analysis [35]. We can adapt our construction to allow downwards and upwards communication of this form, but it is not clear whether Λ remains finite.

Acknowledgement. We thank the anonymous reviewers for their remarks. This work was supported by the Engineering and Physical Sciences Research Council [EP/K009907/1].

References

1. Bouajjani, A., Esparza, J., Maler, O.: Reachability analysis of pushdown automata: application to model-checking. In: Mazurkiewicz, A., Winkowski, J. (eds.) CONCUR 1997. LNCS, vol. 1243, pp. 135–150. Springer, Heidelberg (1997). https://doi.org/10.1007/3-540-63141-0_10
2. Bouajjani, A., Müller-Olm, M., Touili, T.: Regular symbolic analysis of dynamic networks of pushdown systems. In: Abadi, M., de Alfaro, L. (eds.) CONCUR 2005. LNCS, vol. 3653, pp. 473–487. Springer, Heidelberg (2005). https://doi.org/10.1007/11539452_36

3. Brainerd, W.S.: Tree generating regular systems. Inf. Control **14**(2), 217–231 (1969)

4. Broadbent, C.H., Carayol, A., Hague, M., Serre, O.: A saturation method for collapsible pushdown systems. In: Czumaj, A., Mehlhorn, K., Pitts, A., Wattenhofer, R. (eds.) ICALP 2012. LNCS, vol. 7392, pp. 165–176. Springer, Heidelberg (2012). https://doi.org/10.1007/978-3-642-31585-5_18

5. Broadbent, C.H., Carayol, A., Hague, M., Serre, O.: C-SHORe: a collapsible approach to higher-order verification. In: ICFP (2013)

6. Broadbent, C.H., Kobayashi, N.: Saturation-based model checking of higher-order recursion schemes. In: CSL (2013)

7. Chadha, R., Viswanathan, M.: Decidability results for well-structured transition systems with auxiliary storage. In: Caires, L., Vasconcelos, V.T. (eds.) CONCUR 2007. LNCS, vol. 4703, pp. 136–150. Springer, Heidelberg (2007). https://doi.org/10.1007/978-3-540-74407-8_10

8. Clemente, L., Parys, P., Salvati, S., Walukiewicz, I.: Ordered tree-pushdown systems. In: FSTTCS (2015)

9. Clemente, L., Parys, P., Salvati, S., Walukiewicz, I.: The diagonal problem for higher-order recursive schemes is decidable. In: LICS (2016)

10. Cyriac, A., Gastin, P., Kumar, K.N.: MSO decidability of multi-pushdown systems via split-width. In: Koutny, M., Ulidowski, I. (eds.) CONCUR 2012. LNCS, vol. 7454, pp. 547–561. Springer, Heidelberg (2012). https://doi.org/10.1007/978-3-642-32940-1_38

11. Gawlitza, T.M., Lammich, P., Müller-Olm, M., Seidl, H., Wenner, A.: Join-lock-sensitive forward reachability analysis for concurrent programs with dynamic process creation. In: Jhala, R., Schmidt, D. (eds.) VMCAI 2011. LNCS, vol. 6538, pp. 199–213. Springer, Heidelberg (2011). https://doi.org/10.1007/978-3-642-18275-4_15

12. Hague, M.: Saturation of concurrent collapsible pushdown systems. In: FSTTCS (2013)

13. Hague, M., Kochems, J., Ong, C.-H.L.: Unboundedness and downward closures of higher-order pushdown automata. In: POPL (2016)

14. Hague, M., Murawski, A.S., Ong, C.-H.L., Serre, O.: Collapsible pushdown automata and recursion schemes. In: LICS (2008)

15. Hague, M., Ong, C.-H.L.: Winning regions of pushdown parity games: a saturation method. In: Bravetti, M., Zavattaro, G. (eds.) CONCUR 2009. LNCS, vol. 5710, pp. 384–398. Springer, Heidelberg (2009). https://doi.org/10.1007/978-3-642-04081-8_26

16. Hague, M., Penelle, V.: Constrained dynamic tree networks (2018). https://doi.org/10.17637/rh.6850508, https://figshare.com/articles/main_pdf/6850508

17. Kobayashi, N.: Model-checking higher-order functions. In: PPDP (2009)

18. Kobayashi, N.: Higher-order model checking: from theory to practice. In: LICS (2011)

19. Kobayashi, N.: A practical linear time algorithm for trivial automata model checking of higher-order recursion schemes. In: Hofmann, M. (ed.) FoSSaCS 2011. LNCS, vol. 6604, pp. 260–274. Springer, Heidelberg (2011). https://doi.org/10.1007/978-3-642-19805-2_18

20. Kobayashi, N., Igarashi, A.: Model-checking higher-order programs with recursive types. In: Felleisen, M., Gardner, P. (eds.) ESOP 2013. LNCS, vol. 7792, pp. 431–450. Springer, Heidelberg (2013). https://doi.org/10.1007/978-3-642-37036-6_24

21. Kobayashi, N.: GTRecS2: a model checker for recursion schemes based on games and types (2012). http://www-kb.is.s.u-tokyo.ac.jp/~koba/gtrecs2/

22. La Torre, S., Muscholl, A., Walukiewicz, I.: Safety of parametrized asynchronous shared-memory systems is almost always decidable. In: CONCUR (2015)
23. Lammich, P., Müller-Olm, M., Wenner, A.: Predecessor sets of dynamic pushdown networks with tree-regular constraints. In: Bouajjani, A., Maler, O. (eds.) CAV 2009. LNCS, vol. 5643, pp. 525–539. Springer, Heidelberg (2009). https://doi.org/10.1007/978-3-642-02658-4_39
24. Lammich, P., Müller-Olm, M., Seidl, H., Wenner, A.: Contextual locking for dynamic pushdown networks. In: Logozzo, F., Fähndrich, M. (eds.) SAS 2013. LNCS, vol. 7935, pp. 477–498. Springer, Heidelberg (2013). https://doi.org/10.1007/978-3-642-38856-9_25
25. Löding, C.: Infinite graphs generated by tree rewriting. Ph.D. thesis, RWTH Aachen (2003)
26. Lugiez, D.: Forward analysis of dynamic network of pushdown systems is easier without order. Int. J. Found. Comput. Sci. **22**(4), 843–862 (2011)
27. Lugiez, D., Schnoebelen, P.: The regular viewpoint on PA-processes. In: Sangiorgi, D., de Simone, R. (eds.) CONCUR 1998. LNCS, vol. 1466, pp. 50–66. Springer, Heidelberg (1998). https://doi.org/10.1007/BFb0055615
28. Madhusudan, P., Parlato, G.: The tree width of auxiliary storage. In: POPL (2011)
29. Muscholl, A., Seidl, H., Walukiewicz, I.: Reachability for dynamic parametric processes. In: Bouajjani, A., Monniaux, D. (eds.) VMCAI 2017. LNCS, vol. 10145, pp. 424–441. Springer, Cham (2017). https://doi.org/10.1007/978-3-319-52234-0_23
30. Neatherway, R.P., Ramsay, S.J., Ong, C.-H.L.: A traversal-based algorithm for higher-order model checking. In: ICFP (2012)
31. Nordhoff, B., Müller-Olm, M., Lammich, P.: Iterable forward reachability analysis of monitor-DPNs. In: Semantics, Abstract Interpretation, and Reasoning About Programs: Essays Dedicated to David A. Schmidt on the Occasion of his Sixtieth Birthday (2013)
32. Ong, C.-H.L.: On model-checking trees generated by higher-order recursion schemes. In: LICS (2006)
33. Parys, P.: The complexity of the diagonal problem for recursion schemes. In: FSTTCS (2018)
34. Penelle, V.: Rewriting higher-order stack trees. In: Beklemishev, L.D., Musatov, D.V. (eds.) CSR 2015. LNCS, vol. 9139, pp. 364–397. Springer, Cham (2015). https://doi.org/10.1007/978-3-319-20297-6_24
35. Qadeer, S., Rehof, J.: Context-bounded model checking of concurrent software. In: Halbwachs, N., Zuck, L.D. (eds.) TACAS 2005. LNCS, vol. 3440, pp. 93–107. Springer, Heidelberg (2005). https://doi.org/10.1007/978-3-540-31980-1_7
36. Ramsay, S.J., Neatherway, R.P., Ong, C.-H.L.: A type-directed abstraction refinement approach to higher-order model checking. In: POPL (2014)
37. Schwoon, S.: Model-checking pushdown systems. Ph.D. thesis, Technical University of Munich (2002)
38. Seth, A.: Games on higher order multi-stack pushdown systems. In: Bournez, O., Potapov, I. (eds.) RP 2009. LNCS, vol. 5797, pp. 203–216. Springer, Heidelberg (2009). https://doi.org/10.1007/978-3-642-04420-5_19
39. Song, F., Touili, T.: Model checking dynamic pushdown networks. Form. Asp. Comput. **27**(2), 397–421 (2015)
40. Suwimonteerabuth, D., Berger, F., Schwoon, S., Esparza, J.: jMoped: a test environment for java programs. In: Damm, W., Hermanns, H. (eds.) CAV 2007. LNCS, vol. 4590, pp. 164–167. Springer, Heidelberg (2007). https://doi.org/10.1007/978-3-540-73368-3_19

41. Touili, T., Atig, M.F.: Verifying parallel programs with dynamic communication structures. Theor. Comput. Sci. **411**(38–39), 3460–3468 (2010)
42. Yasukata, K., Kobayashi, N., Matsuda, K.: Pairwise reachability analysis for higher order concurrent programs by higher-order model checking. In: Baldan, P., Gorla, D. (eds.) CONCUR 2014. LNCS, vol. 8704, pp. 312–326. Springer, Heidelberg (2014). https://doi.org/10.1007/978-3-662-44584-6_22
43. Yasukata, K., Tsukada, T., Kobayashi, N.: Verification of higher-order concurrent programs with dynamic resource creation. In: Igarashi, A. (ed.) APLAS 2016. LNCS, vol. 10017, pp. 335–353. Springer, Cham (2016). https://doi.org/10.1007/978-3-319-47958-3_18
44. Zetzsche, G.: An approach to computing downward closures. In: Halldórsson, M., Iwama, K., Kobayashi, N., Speckmann, B. (eds.) ICALP 2015. LNCS, vol. 9135, pp. 440–451. Springer, Heidelberg (2015). https://doi.org/10.1007/978-3-662-47666-6_35

EXPSPACE-Complete Variant
of Countdown Games, and Simulation
on Succinct One-Counter Nets

Petr Jančar[1]([⊠]), Petr Osička[1], and Zdeněk Sawa[2]

[1] Department of Computer Science, Faculty of Science,
Palacký University Olomouc, Olomouc, Czechia
`petr.jancar@upol.cz`, `osicka@acm.org`
[2] Department of Computer Science, FEI, Technical University of Ostrava,
Ostrava, Czechia
`zdenek.sawa@vsb.cz`

Abstract. We answer an open complexity question for simulation pre-order of succinct one-counter nets (i.e., one-counter automata with no zero tests where counter increments and decrements are integers written in binary), by showing that all relations between bisimulation equivalence and simulation preorder are EXPSPACE-hard for these nets. We describe a reduction from reachability games whose EXPSPACE-completeness in the case of succinct one-counter nets was shown by Hunter [RP 2015], by using other results. We also provide a direct self-contained EXPSPACE-completeness proof for a special case of such reachability games, namely for a modification of countdown games that were shown EXPTIME-complete by Jurdzinski, Sproston, Laroussinie (LMCS 2008); in our modification the initial counter value is not given but is freely chosen by the first player.

Keywords: Succinct one-counter net · Simulation
Countdown game · Complexity

1 Introduction

One-counter automata (OCA), i.e., finite automata equipped with a nonnegative counter, are studied as one of the simplest models of infinite-state systems. They can be viewed as a special case of Minsky counter machines, or as a special case of pushdown automata. In general, OCA can test the value of the counter for zero, i.e., some transitions could be enabled only if the value of the counter is zero. One-counter nets (OCN) are a "monotonic" subclass of OCA where every transition enabled for zero is also enabled for nonzero values. As usual,

---

This research has been supported by Grant No. 18-11193S, Grant Agency of the Czech Rep. (P. Jančar and P. Osička), and by Grant No. SP2018/172, VŠB-Techn. Univ. Ostrava (Z. Sawa).

ⓒ Springer Nature Switzerland AG 2018
I. Potapov and P.-A. Reynier (Eds.): RP 2018, LNCS 11123, pp. 59–74, 2018.
https://doi.org/10.1007/978-3-030-00250-3_5

we can consider deterministic, nondeterministic, and/or alternating versions of OCA and/or OCN. The basic versions are *unary*, where the counter can be incremented and decremented by one in one step, while in the *succinct* versions the possible changes can be arbitrary integers (but fixed for a given transition); as usual, the changes are assumed to be written in binary in a description of a given automaton.

Problems that have been studied on OCA and OCN include reachability, equivalence, model checking, and also different kinds of games played on these automata. One of the earliest results showed decidability of (language) equivalence for deterministic OCA [24]. (The open polynomiality question in [24] was positively answered in [2].)

Later other behavioural equivalences (besides language equivalence) have been studied. Most relevant for us is the research started by Abdulla and Čerāns who showed in [1] that simulation preorder on one-counter nets is decidable. An alternative proof of this fact was given in [14]; it was also noted that simulation equivalence is undecidable for OCA. A relation to bisimulation problems was shown in [12]. Kučera showed some lower bounds in [19]; Mayr [20] showed the undecidability of weak bisimulation equivalence on OCN.

Simulation preorder on one-counter nets turned out PSPACE-complete: the lower bound was shown by Srba [23], and the upper bound by Hofman, Lasota, Mayr, and Totzke [10]. It was also shown in [10] that deciding weak simulation on OCN can be reduced to deciding strong simulation on OCN, and thus also solved in polynomial space. (Strong) bisimulation equivalence on OCA is also known to be PSPACE-complete [3].

The mentioned results deal with unary OCA and OCN. Succinct (and parametric) OCA were considered, e.g., in [9], where reachability on succinct OCA was shown to be NP-complete. We note that PSPACE-membership of problems for the unary case easily yields EXPSPACE-membership for the succinct (binary) case. Games studied on OCA include, e.g., parity games on one-counter processes (with test for zero) [22], and are closely related to counter reachability games (e.g. [21]). Model checking problems on OCA were studied for many types of logics, e.g., LTL [5], branching time logics [7], or first-order logics [8]. DP-lower bounds for some model-checking (and also equivalence checking) problems were shown in [13].

An involved result by Göller, Haase, Ouaknine, Worrell [6] shows that model checking a fixed CTL formula on succinct one-counter automata is EXPSPACE-hard. Their proof is interesting and nontrivial, and uses two involved results from complexity theory. The technique of this proof was referred to by Hunter [11], to derive EXPSPACE-hardness of reachability games on succinct one-counter nets.

Our Contribution. In this paper we close a complexity gap for the simulation problem that was mentioned in [10], noting that there was a PSPACE lower bound and an EXPSPACE upper bound for the problem. We show EXPSPACE-hardness (and thus EXPSPACE-completeness) of the problem, using a defender-choice technique (cf., e.g., [16]) to reduce reachability games to any relation between simulation preorder and bisimulation equivalence.

The EXPSPACE-hardness can be derived by [6,11]. Here we present a direct proof of EXPSPACE-hardness (and completeness) of a special case of reachability games, which we call the "existential countdown games". It is a mild relaxation of the countdown games from [17] (or their variant from [18]) which is an interesting EXPTIME-complete problem. We thus provide a complete EXPSPACE-hardness proof, independent of [11] or [6].

Organization of the Paper. Section 2 gives the basic definitions. In Sect. 3 we show the "existential" countdown games and their EXPSPACE-completeness. Section 4 describes the reductions from reachability games to (bi)simulation relations. We finish with some additional remarks in Sect. 5.

2 Basic Definitions

By $\mathbb{Z}$ and $\mathbb{N}$ we denote the sets of integers and of nonnegative integers, respectively. We use $[i,j]$, where $i,j \in \mathbb{Z}$, for denoting the set $\{i, i+1, \ldots, j\}$ (which is empty when $i > j$).

Labelled Transition Systems and (Bi)simulations. A *labelled transition system*, an *LTS* for short, is a tuple

$$\mathcal{L} = (S, Act, (\xrightarrow{a})_{a \in Act})$$

where S is the set of *states*, Act is the set of *actions*, and $\xrightarrow{a} \subseteq S \times S$ is the set of *a-transitions* (transitions labelled with a), for each $a \in Act$. We write $s \xrightarrow{a} t$ instead of $(s,t) \in \xrightarrow{a}$. By $s \xrightarrow{a}$ we denote that a is *enabled* in s, i.e., $s \xrightarrow{a} t$ for some t.

Given $\mathcal{L} = (S, Act, (\xrightarrow{a})_{a \in Act})$, a relation $R \subseteq S \times S$ is a *simulation* if for every $(s,s') \in R$ and every $s \xrightarrow{a} t$ there is $s' \xrightarrow{a} t'$ such that $(t,t') \in R$; if, moreover, for every $(s,s') \in R$ and every $s' \xrightarrow{a} t'$ there is $s \xrightarrow{a} t$ such that $(t,t') \in R$, then R is a *bisimulation*.

The union of all simulations (on S) is the maximal simulation, denoted $\preceq$; it is a preorder, called *simulation preorder*. The union of all bisimulations is the maximal bisimulation, denoted $\sim$; it is an equivalence, called *bisimulation equivalence* (or *bisimilarity*). We obviously have $\sim \subseteq \preceq$.

(Labelled) One-Counter Nets (OCNs and SOCNs). A *labelled one-counter net*, or just a *one-counter net* or even just an *OCN* for short, is a triple

$$\mathcal{N} = (Q, Act, \delta),$$

where Q is the finite set of *control states*, Act the finite set of *actions*, and $\delta \subseteq Q \times Act \times \{-1, 0, +1\} \times Q$ is the finite set of *(labelled transition) rules*. By allowing $\delta \subseteq Q \times Act \times \mathbb{Z} \times Q$, and presenting $z \in \mathbb{Z}$ in the rules (q, a, z, q') in binary, we get a *succinct one-counter net*, or a *SOCN* for short. (One-counter automaton arises by adding the ability to test explicitly if the counter is zero.) We present rules (q, a, z, q') rather as $q \xrightarrow{a,z} q'$.

Each OCN or SOCN $\mathcal{N} = (Q, Act, \delta)$ has the *associated LTS*

$$\mathcal{L}_\mathcal{N} = (Q \times \mathbb{N}, Act, (\xrightarrow{a})_{a \in Act}) \tag{1}$$

where $(q, m) \xrightarrow{a} (q', n)$ iff $q \xrightarrow{a, n-m} q'$ is a rule in δ. We often write a state (q, m), which is also called a *configuration*, in the form $q(m)$, and we view m as a value of a nonnegative counter. A rule $q \xrightarrow{a,z} q'$ thus induces transitions $q(m) \xrightarrow{a} q'(m + z)$ for all $m \geq \max\{0, -z\}$.

Reachability Games (r-Games), Winning Areas, Ranks of States. By a *reachability game*, or an *r-game* for short, we mean a tuple

$$\mathcal{G} = (V, V_\exists, \rightarrow, \mathcal{T}),$$

where V is the set of *states* (or *vertices*), $V_\exists \subseteq V$ is the set of *Eve's states*, $\rightarrow \subseteq V \times V$ is the *transition relation* (or the set of *transitions*), and $\mathcal{T} \subseteq V$ is the set of *target states*. By *Adam's states* we mean the elements of $V_\forall = V \setminus V_\exists$.

Eve's winning area is $Win_\exists = \bigcup_{\lambda \in Ord} W_\lambda$, for Ord being the class of ordinals, where the sets $W_\lambda \subseteq V$ are defined inductively as follows.

We put $W_0 = \mathcal{T}$; for $\lambda > 0$ we put $W_{<\lambda} = \bigcup_{\lambda' < \lambda} W_{\lambda'}$, and we stipulate:

(a) if $s \notin W_{<\lambda}$, $s \in V_\exists$, and $s \rightarrow \bar{s}$ for some $\bar{s} \in W_{<\lambda}$, then $s \in W_\lambda$;
(b) if $s \notin W_{<\lambda}$, $s \in V_\forall$, and we have $\emptyset \neq \{\bar{s} \mid s \rightarrow \bar{s}\} \subseteq W_{<\lambda}$, then $s \in W_\lambda$.

(If (a) applies, then λ is surely a successor ordinal.)

For each $s \in Win_\exists$, by RANK(s) we denote (the unique) λ such that $s \in W_\lambda$. A transition $s \rightarrow \bar{s}$ is *rank-reducing* if RANK$(s) >$ RANK$(\bar{s})$. We note that for any $s \in Win_\exists$ with RANK$(s) > 0$ we have: if $s \in V_\exists$, then there is at least one rank-reducing transition $s \rightarrow \bar{s}$ (in fact, RANK$(s) =$ RANK$(\bar{s}) + 1$ in this case); if $s \in V_\forall$, then there is at least one transition $s \rightarrow \bar{s}$ and all such transitions are rank-reducing. This entails that $Win_\exists$ is the set of states from which Eve has a winning strategy that guarantees reaching (some state in) $\mathcal{T}$ when Eve is choosing a next transition in Eve's states and Adam is choosing a next transition in Adam's states.

We are primarily interested in the games that have (at most) countably many states and are finitely branching (the sets $\{\bar{s} \mid s \rightarrow \bar{s}\}$ are finite for all s). In such cases we have RANK$(s) \in \mathbb{N}$ for each $s \in Win_\exists$.

We now define specific r-games, presented by (unlabelled) SOCNs with partitioned control-state sets.

Succinct One-Counter Net r-Games (socn-r-Games), Problem SOC-NRG. By a *succinct one-counter net r-game*, a *socn-r-game* for short, we mean a tuple

$$\mathcal{N} = (Q, Q_\exists, \delta, p_{win})$$

where Q is the finite set of *(control) states*, $Q_\exists \subseteq Q$ is the set of *Eve's (control) states*, $p_{win} \in Q$ is the *target (control) state*, and $\delta \subseteq Q \times \mathbb{Z} \times Q$ is the finite set of *(transition) rules*. We often present a rule $(q, z, q') \in \delta$ as $q \xrightarrow{z} q'$. By

Adam's (control) states we mean the elements of $Q_\forall = Q \setminus Q_\exists$. A socn-r-game $\mathcal{N} = (Q, Q_\exists, \delta, p_{win})$ has the *associated r-game*

$$\mathcal{G}_\mathcal{N} = (Q \times \mathbb{N}, Q_\exists \times \mathbb{N}, \rightarrow, \{(p_{win}, 0)\}) \tag{2}$$

where $(q, m) \rightarrow (q', n)$ iff $q \xrightarrow{n-m} q'$ is a rule (in δ). We often write $q(m)$ instead of (q, m) for states of $\mathcal{G}_\mathcal{N}$. We define the problem *SOCNRG* (to decide succinct one-counter net r-games) as follows:

Instance: a socn-r-game $\mathcal{N}$ (with integers z in rules $q \xrightarrow{z} q'$ written in binary), and a control state p_0.
Question: is $p_0(0) \in Win_\exists$ in the game $\mathcal{G}_\mathcal{N}$?

Remark. We have defined the target states (in $\mathcal{G}_\mathcal{N}$) to be the singleton set $\{p_{win}(0)\}$. There are other natural variants (e.g., one in [11] defines the target set $\{p(0) \mid p \neq p_0\}$) that can be easily shown to be essentially equivalent.

3   EXPSPACE-Completeness of Existential Countdown Games

The EXPSPACE-hardness of SOCNRG was announced in [11], where an idea of a proof is sketched, also using a reference to an involved result [6] (which is further discussed in Sect. 5). Here we give a direct self-contained proof that does not rely on [11] or involved techniques from [6], and that even shows that SOCNRG is EXPSPACE-hard already in the special case that slightly generalizes the countdown games from [17]. (The EXPSPACE-membership follows from [11], but we add a short proof to be self-contained.)

We define a *countdown game* as a socn-r-game $\mathcal{N} = (Q, Q_\exists, \delta, p_{win})$, where in every rule $q \xrightarrow{z} q'$ in δ we have $z < 0$. The problem CG is defined as follows:

Instance: a countdown game $\mathcal{N}$ (with integers in rules written in binary), and an initial configuration $p_0(n)$ where $n \in \mathbb{N}$ (n in binary).
Question: is $p_0(n) \in Win_\exists$?

The problem CG (in an equivalent form) was shown EXPTIME-complete in [17]. Here we define an existential version, i.e. the problem ECG:

Instance: a countdown game $\mathcal{N}$ and a control state p_0.
Question: is there some $n \in \mathbb{N}$ such that $p_0(n) \in Win_\exists$?

ECG can be indeed viewed as a subproblem of SOCNRG: given an instance of ECG, it suffices to add a new Eve's state p'_0 and rules $p'_0 \xrightarrow{1} p'_0$, $p'_0 \xrightarrow{0} p_0$; the question then is if $p'_0(0) \in Win_\exists$.

In the rest of this section we prove the following theorem.

Theorem 1. *ECG (existential countdown games) is EXPSPACE-complete.*

EXPSPACE-Hardness of ECG. We use a "master" reduction. We thus fix an arbitrary language $L \subseteq \Sigma^*$ in EXPSPACE, decided by a (deterministic) Turing machine M in space $2^{p(n)}$ for a fixed polynomial p. For any word $w \in \Sigma^*$ there is the respective computation of M using at most $m = 2^{p(n)}$ tape cells (where $n = |w|$), which is accepting iff $w \in L$. We show a construction of a countdown game $\mathcal{N}_{w,m}^M$ with an initial control state p_0 such that there exists $k \geq 0$ where $p(k) \in Win_\exists$ if, and only if, M accepts w. Moreover, for the fixed M, logarithmic space (with respect to n) is sufficient to construct $\mathcal{N}_{w,m}^M$. Thus, there is a logspace reduction from the membership problem for L to the problem ECG.

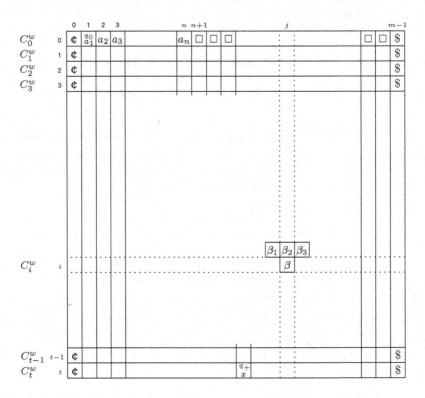

Fig. 1. A table of configurations in the computation of M on word $w = a_1 a_2 \ldots a_n$

Construction Informally. The construction of the countdown game $\mathcal{N}_{w,m}^M$ elaborates an idea that is already present in [4] (in Theorem 3.4) and that was also used, e.g., in [15]. We first present the game informally; this is then formalized in a straightforward technical way.

Figure 1 presents an accepting computation of M, on a word $w = a_1 a_2 \ldots a_n$; it starts in the initial control state q_0 with the head scanning a_1. The computation is a sequence of configurations $C_0^w, C_1^w, \ldots, C_t^w$, where C_t^w is accepting (since the control state is q_+). We assume that M never leaves its input to the left, hence the tape position 0, which is filled with a special "left sentinel" ¢ for transparency, is never visited. On the other hand, the position $m - 1$ of the right sentinel $ is never visited either; the space complexity of the computation is thus at most $m - 2$.

Let us imagine a game where Eve, given w, claims that w is accepted by M. Nevertheless she does not present a respective accepting computation; she only produces a "row" $r \in \mathbb{N}$ and a "column" $c \in \mathbb{N}$, and a tape-symbol x, claiming that if we constructed the computation $C_0^w, C_1^w, \ldots, C_t^w$, then we would find that ($r = t$ and) C_r^w is the accepting configuration and the symbol on position c in C_r^w is (q_+, x) (i.e., the tape symbol is x, and the head happens to scan position c, and the control state is q_+).

Generally, if Eve claims that in the row i and the column j we would find β if we constructed the computation (where β is either a tape symbol or a tape symbol combined with a control state), she must present a triple $(\beta_1, \beta_2, \beta_3)$ in the previous row $i - 1$ as depicted in Fig. 1, and Adam chooses one of symbols $\beta_1, \beta_2, \beta_3$ for the next round; the triple must be consistent with β w.r.t. the rules of M. If in the row 0 Adam chooses a symbol that is correct in C_0^w, then Eve wins (otherwise Adam wins). It is easy to verify that Eve has a winning strategy in this game iff w is accepted by M.

We note that the described game uses a pair of number-variables i, j, to determine the current cell in the computation table. But we can ask Eve to provide some $m \in \mathbb{N}$ in the beginning (claiming that the head only moves between positions 1 and $m-2$ during the computation); the pair (i, j) can be then represented by the value $z = i \cdot m + j$ (and going from i to $i-1$ amounts to subtract m from z). If Eve sometimes claims that $\beta_1 = $ ¢, then the respective column-position is 0, which entails that the respective value $z = i \cdot m + j$ should be divisible by m; if Adam doubts this, it can be verified by subtracting m repeatedly. Similarly, if Eve claims $\beta_3 = $ \$, then by subtracting $m - 1$ we should get a number divisible by m. If Eve claims something else than ¢ in a position corresponding to the ¢-column, or something else than \$ in a position corresponding to the \$-column, then Adam just keeps choosing this column, and Eve's cheating is revealed in the row 0.

Construction Formally. Now we formalize the above idea, which is a routine technical work, in fact.

Assume a fixed deterministic Turing machine $M = (Q, \Sigma, \Gamma, \delta, q_0, \{q_+, q_-\})$, where Q is the set of (control) states, $q_0 \in Q$ the initial state, $q_+ \in Q$ the accepting state, $q_- \in Q$ the rejecting state, Σ the input alphabet, $\Gamma \supseteq \Sigma$ the tape alphabet, satisfying $\square \in \Gamma \smallsetminus \Sigma$ for the special blank tape-symbol $\square$, and $\delta : (Q \smallsetminus \{q_+, q_-\}) \times \Gamma \to Q \times \Gamma \times \{-1, +1\}$ is the transition function.

Putting $\Delta = \Gamma \cup (Q \times \Gamma)$, we define the consistency relation $\vdash \subseteq \Delta^3 \times \Delta$ in a standard way: $(\beta_1, \beta_2, \beta_3) \vdash \beta$ (to be read "β is consistent with $(\beta_1, \beta_2, \beta_3)$") if $\beta_i \in Q \times \Gamma$ for at most one $i \in \{1, 2, 3\}$ and the following conditions hold:

- if $\beta_1\beta_2\beta_3 = (q, x)yz$ and $\delta(q, x) = (q', x', d)$, then $\beta = (q', y)$ if $d = +1$ and $\beta = y$ otherwise (i.e., if $d = -1$);

- if $\beta_1\beta_2\beta_3 = x(q,y)z$ and $\delta(q,y) = (q',y',d)$, then $\beta = y'$ (for any $d \in \{-1,+1\}$);
- if $\beta_1\beta_2\beta_3 = xy(q,z)$ and $\delta(q,z) = (q',z',d)$, then $\beta = (q',y)$ if $d = -1$ and $\beta = y$ otherwise;
- if $\beta_1\beta_2\beta_3 = xyz$, then $\beta = y$.

We note that $\vdash$ is a partial function, in fact. By a *configuration* of M we mean a mapping $C : \mathbb{Z} \to \Delta$ where $C(j) \neq \square$ for only finitely many $j \in \mathbb{Z}$ and $C(j) \in Q \times \Gamma$ for precisely one $j \in \mathbb{Z}$, called the *head-position*; if $C(j) = (q_+,x)$ for the head-position j (and $x \in \Gamma$) then C is *accepting*, and if $C(j) = (q_-,x)$ then C is *rejecting*.

We put $C \vdash C'$ (thus overloading the symbol $\vdash$) if $(C(j-1), C(j), C(j+1)) \vdash C'(j)$ for all $j \in \mathbb{Z}$. This relation $\vdash$ is again a partial function; if C is *final*, i.e. accepting or rejecting, then there is no C' such that $C \vdash C'$.

Given a word $w = a_1a_2 \cdots a_n \in \Sigma^*$ (hence $|w| = n$), we define the respective initial configuration as C_0^w where $C_0^w(1) = (q_0,a_1)$ if $n \geq 1$ and $C_0^w(1) = (q_0,\square)$ if $n = 0$, $C_0^w(j) = a_j$ for all $j \in [2,n]$, and $C_0^w(j) = \square$ for all $j \leq 0$ and all $j > n$. If C_i^w is not final, then we define C_{i+1}^w so that $C_i^w \vdash C_{i+1}^w$. The *computation on w* is either the finite sequence $C_0^w, C_1^w, C_2^w, \ldots, C_t^w$ where C_t^w is final (accepting or rejecting), or the infinite sequence $C_0^w, C_1^w, C_2^w, \ldots$; formally we put $C_i^w(j) = \bot$ (for $\bot \notin \Delta$) if there is a final C_t^w and $i > t$.

By $L(M)$ we denote the *language accepted by M*, i.e. the set $\{w \in \Sigma^* \mid$ the computation on w finishes with an accepting configuration$\}$.

Now we assume that the Turing machine M uses a bounded space for the input $w = a_1a_2 \ldots a_n$, and in particular that we have $m \in \mathbb{N}$ such that during the computation of M on w the head-position is never outside $[1,m-2]$; for technical convenience and without loss of generality we also assume that $m \geq n \geq 1$ and $m > 3$. We can imagine that the computation is presented as a table depicted in Fig. 1, and the columns 0 and $m-1$ are filled with special symbols ¢ and \$ (where ¢, \$ $\notin \Delta$), i.e., $C_i^w(0) = $ ¢ and $C_i^w(m-1) = \$$ for each $i \geq 0$. We also extend relation $\vdash$ accordingly to incorporate these special symbols. In particular, whenever $(\beta_1, \beta_2, \beta_3) \vdash \beta$, we have $\beta \in \Delta$, $\beta_1 \in \Delta \cup \{$¢$\}$, $\beta_2 \in \Delta$, and $\beta_3 \in \Delta \cup \{\$\}$, and we exclude those combinations of $\beta_1, \beta_2, \beta_3$ that would correspond to a move of the head of the machine M to a position containing ¢ or \$.

Given a Turing machine M, its input $w = a_1a_2 \cdots a_n$, and a number m, satisfying the above assumptions, we construct a corresponding countdown game

$$\mathcal{N}_{w,m}^M = (\overline{Q}, \overline{Q}_\exists, \delta_\mathcal{N}, p_{win})$$

where

- $\overline{Q}_\exists = \{p_0, p_2, p_{win}, p_{bad}, s_\text{¢}, s_\$\} \cup \{s_\beta \mid \beta \in \Delta\}$,
- $\overline{Q}_\forall = \{p_1\} \cup \{s_{(\beta_1,\beta_2,\beta_3)} \mid \beta_i \in \Delta\} \cup \{s_{(\text{¢},\beta_2,\beta_3)} \mid \beta_i \in \Delta\} \cup \{s_{(\beta_1,\beta_2,\$)} \mid \beta_i \in \Delta\}$

(recall that $\overline{Q}_\forall = \overline{Q} \setminus \overline{Q}_\exists$), and the set $\delta_\mathcal{N}$ consists of the rules in Fig. 2 (for all $\beta \in \Delta$, $\beta_1 \in \Delta \cup \{$¢$\}$, $\beta_2 \in \Delta$, $\beta_3 \in \Delta \cup \{\$\}$).

The relation of the countdown game $\mathcal{N}_{w,m}^M$ to the computation of M on w is stated in the following proposition.

$$p_0 \xrightarrow{-1} s_{(q_+,x)} \qquad \text{(where } x \in \Gamma) \tag{1}$$

$$s_\beta \xrightarrow{-(m-2)} s_{(\beta_1,\beta_2,\beta_3)} \qquad \text{(where } (\beta_1,\beta_2,\beta_3) \vdash \beta) \tag{2}$$

$$s_{(\beta_1,\beta_2,\beta_3)} \xrightarrow{-3} s_{\beta_1} \qquad s_{(\beta_1,\beta_2,\beta_3)} \xrightarrow{-2} s_{\beta_2} \qquad s_{(\beta_1,\beta_2,\beta_3)} \xrightarrow{-1} s_{\beta_3} \tag{3}$$

$$s_\varnothing \xrightarrow{-m} s_\varnothing \qquad s_\varnothing \xrightarrow{-2} p_{win} \qquad s_\$ \xrightarrow{-(m-1)} s_\varnothing \tag{4}$$

$$s_\beta \xrightarrow{-(2+j)} p_{win} \qquad \text{(where } 1 \le j \le n \text{ and } C_0^w(j) = \beta) \tag{5}$$

$$s_\square \xrightarrow{-(n+1)} p_1 \qquad p_1 \xrightarrow{-(m-n)} p_{bad} \qquad p_1 \xrightarrow{-1} p_2 \tag{6}$$

$$p_2 \xrightarrow{-1} p_2 \qquad p_2 \xrightarrow{-1} p_{win}$$

Fig. 2. Rules of $\mathcal{N}_{w,m}^M$

Proposition 2. *For the countdown game $\mathcal{N}_{w,m}^M$ there exists $k \ge 0$ such that $p_0(k) \in Win_\exists$ iff the computation of M on w never moves the head out of $[1, m-2]$ and finishes in an accepting configuration.*

Proof. The configurations of $\mathcal{N}_{w,m}^M$ of the form $s_\beta(k)$ with $\beta \in \Delta \cup \{\varnothing, \$\}$ and $k = 2+i \cdot m+j$ where $i \ge 0$ and $0 \le j < m$ correspond to the situation where Eve claims that in the table of configurations of computation of M on w (see Fig. 1) the cell in row i and column j contains symbol β. We will show that she has a winning strategy from $s_\beta(k)$ exactly when this is the case, i.e., $s_\beta(k) \in Win_\exists$ iff $C_i^w(j) = \beta$. (In our construction, we need to add number 2 to $i \cdot m + j$ to ensure that in every move in the game the value of the counter is decremented by at least 1. It would not be necessary if we allow moves that do not change the counter value.)

To prove that $s_\beta(k) \in Win_\exists$ iff $C_i^w(j) = \beta$ (for $k = 2 + i \cdot m + j$), we start by the following facts that are easy to check (recall that Eve wins iff the configuration $p_{win}(0)$ is reached):

(a) $s_\varnothing(k) \in Win_\exists$ iff $k = 2 + i \cdot m$ for some $i \in \mathbb{N}$;
(b) $s_\$(k) \in Win_\exists$ iff $k = 2 + i \cdot m + (m-1)$ for some $i \in \mathbb{N}$;
(c) for $1 \le j \le n$, $s_\beta(2+j) \in Win_\exists$ iff $\beta = C_0^w(j)$;
(d) $p_1(k) \in Win_\exists$ iff $2 \le k < m - n$;
(e) for $n < j < m - 1$, $s_\beta(2+j) \in Win_\exists$ iff $\beta = \square$.

Assume now that $\beta \in \Delta \cup \{\varnothing, \$\}$, $i \ge 0$, $0 \le j < m$, and $k = 2 + i \cdot m + j$. To show $s_\beta(k) \in Win_\exists$ iff $C_i^w(j) = \beta$, we proceed by induction on i:

- *Base case $i = 0$:* In this case $k = 2 + j$, so we need to show that $s_\beta(2 + j) \in Win_\exists$ iff $C_0^w(j) = \beta$. This follows easily from facts (a), (b), (c), (e) mentioned above because the initial configuration C_0^w consists of symbol $\varnothing$ ($j = 0$, fact (a)), the input word with the initial control state ($1 \le j \le n$, fact (c)), blanks ($n < j < m-1$, fact (e)), and symbol $\$$ ($j = m-1$, fact (b)).
- *Induction step $i > 0$:* If $\beta = \varnothing$, Eve wins iff $j = 0$ (by fact (a)). Similarly, if $\beta = \$$, she wins iff $j = m - 1$ (by fact (b)).
   Assume now that $\beta \in \Delta$. Eve will lose if she uses any rule from groups (5) or (6) in $s_\beta(k)$, as can be easily checked, and she cannot play rules from

group (4), so she is forced to use some rule from group (2). By playing $s_\beta(k) \xrightarrow{m-2} s_{(\beta_1,\beta_2,\beta_3)}(k')$, where $k' = 2 + (i - 1) \cdot m + j + 2$, she chooses a triple $(\beta_1,\beta_2,\beta_3)$ satisfying $(\beta_1,\beta_2,\beta_3) \vdash \beta$, where β_1, β_2, β_3 are symbols supposedly occurring on positions $j - 1$, j, $j + 1$ in configuration C_{i-1}^w. If β is incorrect (i.e., if $C_i^w(j) \neq \beta$), then at least one of β_1, β_2, β_3 must be also incorrect, and if β is correct, then Eve can choose correct β_1, β_2, β_3. Now Adam can challenge some of the symbols β_1, β_2, β_3 by choosing $\ell \in \{1,2,3\}$ and playing $s_{(\beta_1,\beta_2,\beta_3)}(k') \xrightarrow{-(4-\ell)} s_{\beta_\ell}(k'')$, for $k'' = 2+(i-1)\cdot m+j+(\ell-2)$. By the induction hypothesis he thus has a possibility to preclude Eve's win precisely when one of β_1, β_2, β_3 is incorrect. In particular, if $j = 0$ and $\beta \neq \mathfrak{c}$, or if $j = m - 1$ and $\beta \neq \$$, then Adam repeatedly chooses $\ell = 2$, thus staying in the same column, and Eve has no possibility to install the correct $\mathfrak{c}$, resp. $\$$, in this column anymore.

It is now clear that if the computation of M on w never moves the head outside $[1, m - 2]$ and finishes in an accepting configuration C_r^w with $C_r^w(j) = (q_+, x)$ for some $x \in \Gamma$ and $1 \leq j < m - 1$, then Eve can force her win in $p_0(k)$ where $k = 3+r\cdot m+j$ by playing $p_0(k) \xrightarrow{-1} s_{(q_+,x)}(2+r\cdot m+j)$. It is also easy to check that if the computation moves the head outside $[1, m - 2]$ or is not accepting, then Eve cannot force her win from $p_0(k)$ for any $k \in \mathbb{N}$.   □

We note that the control states in $\mathcal{N}_{w,m}^M$ are determined by M. The rules of $\mathcal{N}_{w,m}^M$, except those in group (5), depend only on M and "parameters" $n = |w|$ and m.

To finish the EXPSPACE-hardness part of the proof of Theorem 1, we assume an arbitrary fixed language L in EXPSPACE. There is thus a Turing machine M and a polynomial p such that M accepts L and the head-position in the computation of M on any w (in the alphabet of L) never moves out of the interval $[1, m - 2]$ where $m = 2^{p(n)}$ for $n = |w|$. Given w, it is straightforward to construct $\mathcal{N}_{w,m}^M$, by filling the parameters n, m, and the rules in group (5), into a fixed scheme. Since m can be presented in binary by using $p(n) + 1$ bits, we can construct $\mathcal{N}_{w,m}^M$ in logarithmic work-space (from a given w).

ECG Is in EXPSPACE. Given an ECG-instance $\mathcal{N} = (Q, Q_\exists, \delta, p_{win})$, p_0, we can stepwise construct $W(0), W(1), W(2), \ldots$ where $W(j) = (Q \times \{j\}) \cap Win_\exists$. For determining $W(n)$ it suffices to know the segment $W(n - \text{M}), W(n - \text{M} + 1)$, $\ldots, W(n - 1)$ where M is the maximum value by which the counter can be decremented in one step. By the pigeon-hole principle, such exponential-size segments must repeat inside $W(0), W(1), \ldots, W(2^{|Q|\text{M}})$. Hence if there is n such that $p_0(n) \in Win_\exists$, then there is such n of double-exponential size; exponential space is thus sufficient for finding such n.

4   Reachability Game Reduces to (Bi)simulation Game

We show a reduction for general r-games, and then apply it to the case of socn-r-games. This yields a log-space reduction of SOCNRG to behavioural relations.

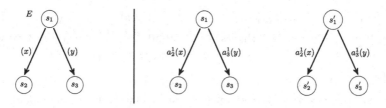

Fig. 3. Eve's state s_1 in $\mathcal{G}$ (left) is mimicked by the pair (s_1, s_1') in $\mathcal{L}(\mathcal{G})$ (right); it is thus Attacker who chooses (s_2, s_2') or (s_3, s_3') as the next current pair.

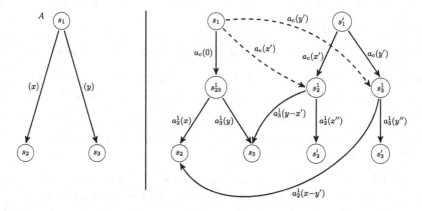

Fig. 4. In (s_1, s_1') it is, in fact, Defender who chooses (s_2, s_2') or (s_3, s_3') (when Attacker avoids pairs with equal states); to take the counter-changes into account correctly, we put $x' = \min\{x, 0\}$, $x'' = \max\{x, 0\}$, and $y' = \min\{y, 0\}$, $y'' = \max\{y, 0\}$ (hence $x = x' + x''$ and $y = y' + y''$). (The dashed edges are viewed as the other edges.)

4.1   Reduction in a General Framework

We assume an r-game $\mathcal{G}$, and below we define a "mimicking" LTS $\mathcal{L}(\mathcal{G})$. In illustrating Figs. 3 and 4 we now ignore the bracketed parts of transition-labels; hence, e.g., in Fig. 3 we can see the transition $s_1 \to s_2$ in $\mathcal{G}$ on the left and the (corresponding) transitions $s_1 \xrightarrow{a_2^1} s_2$ and $s_1' \xrightarrow{a_2^1} s_2'$ in $\mathcal{L}(\mathcal{G})$ on the right. We also use an informal (bi)simulation game terminology: in a current pair of states (e.g., in (s_1, s_1') in $\mathcal{L}(\mathcal{G})$), Attacker performs a transition on one side, and Defender responds with a "same-label" transition on the other side, which yields a new current pair; in the bisimulation game Attacker chooses the sides freely, in the simulation game he must always choose the left-hand side.

So let $\mathcal{G} = (V, V_\exists, \to, \mathcal{T})$ be an r-game, where $V_\forall = V \setminus V_\exists$; we define $\mathcal{L}(\mathcal{G}) = (S, Act, (\xrightarrow{a})_{a \in Act})$ as follows. We put

$$S = V \cup V' \cup \{\langle s, \bar{s}\rangle \mid s \in V_\forall, s \to \bar{s}\} \cup \{\langle s, X\rangle \mid s \in V_\forall, X = \{\bar{s} \mid s \to \bar{s}\} \neq \emptyset\}$$

where $V' = \{s' \mid s \in V\}$ is a "copy" of V. (In Fig. 4 we write, e.g., s_3^1 instead of $\langle s_1, s_3\rangle$, and s_{23}^1 instead of $\langle s_1, \{s_2, s_3\}\rangle$.)

We put $Act = \{a_c, a_{win}\} \cup \{a_{\langle s, \bar{s} \rangle} \mid s \to \bar{s}\}$ and define $\xrightarrow{a}$ for $a \in Act$ as follows. If $s \in V_\exists$ and $s \to \bar{s}$, then $s \xrightarrow{a_{\langle s, \bar{s} \rangle}} \bar{s}$ and $s' \xrightarrow{a_{\langle s, \bar{s} \rangle}} \bar{s}'$ (in Fig. 3 we write, e.g., a_3^1 instead of $a_{\langle s_1, s_3 \rangle}$). If $s \in V_\forall$ and $X = \{\bar{s} \mid s \to \bar{s}\} \neq \emptyset$, then:

(a) $s \xrightarrow{a_c} \langle s, X \rangle$, and $s \xrightarrow{a_c} \langle s, \bar{s} \rangle$, $s' \xrightarrow{a_c} \langle s, \bar{s} \rangle$ for all $\bar{s} \in X$ (cf. Fig. 4 where $s = s_1$ and $X = \{s_2, s_3\}$ and consider dashed edges as normal edges; a_c is a "choice-action");

(b) for each $\bar{s} \in X$ we have $\langle s, X \rangle \xrightarrow{a_{\langle s, \bar{s} \rangle}} \bar{s}$ and $\langle s, \bar{s} \rangle \xrightarrow{a_{\langle s, \bar{s} \rangle}} \bar{s}'$; moreover, for each $\bar{\bar{s}} \in X \smallsetminus \{\bar{s}\}$ we have $\langle s, \bar{s} \rangle \xrightarrow{a_{\langle s, \bar{s} \rangle}} \bar{\bar{s}}$ (e.g., in Fig. 4 we thus have $s_2^1 \xrightarrow{a_2^1} s_2'$ and $s_2^1 \xrightarrow{a_3^1} s_3$).

For each $s \in T$ we have $s \xrightarrow{a_{win}} s$ (for special a_{win} that is not enabled in s').

Lemma 3. *For an r-game $\mathcal{G} = (V, V_\exists, \to, T)$ and its "mimicking" LTS $\mathcal{L}(\mathcal{G}) = (S, Act, (\xrightarrow{a})_{a \in Act})$, the following conditions hold for every $s \in V$ and every relation ρ satisfying $\sim \, \subseteq \rho \subseteq \, \preceq$:*

(a) if $s \in Win_\exists$ (in $\mathcal{G}$), then $s \npreceq s'$ (in $\mathcal{L}(\mathcal{G})$) and thus $(s, s') \notin \rho$;
(b) if $s \notin Win_\exists$, then $s \sim s'$ and thus $(s, s') \in \rho$.

Proof. (a) For the sake of contradiction suppose that there is $s \in Win_\exists$ such that $s \preceq s'$; we consider such s with the least rank. We note that $\text{RANK}(s) > 0$, since $s \in T$ entails $s \npreceq s'$ due to the transition $s \xrightarrow{a_{win}} s$. If $s \in V_\exists$, then let $s \to \bar{s}$ be a rank-reducing transition. Attacker's move $s \xrightarrow{a_{\langle s, \bar{s} \rangle}} \bar{s}$, from the pair (s, s'), must be responded with $s' \xrightarrow{a_{\langle s, \bar{s} \rangle}} \bar{s}'$; but we have $\bar{s} \npreceq \bar{s}'$ by the "least-rank" assumption, which contradicts with the assumption $s \preceq s'$. If $s \in V_\forall$, then $X = \{\bar{s} \mid s \to \bar{s}\}$ is nonempty (since $s \in Win_\exists$) and $\text{RANK}(\bar{s}) < \text{RANK}(s)$ for all $\bar{s} \in X$. For the pair (s, s') we now consider Attacker's move $s \xrightarrow{a_c} \langle s, X \rangle$. Defender can choose $s' \xrightarrow{a_c} \langle s, \bar{s} \rangle$ for any $\bar{s} \in X$ (recall that $\text{RANK}(\bar{s}) < \text{RANK}(s)$). In the current pair $(\langle s, X \rangle, \langle s, \bar{s} \rangle)$ Attacker can play $\langle s, X \rangle \xrightarrow{a_{\langle s, \bar{s} \rangle}} \bar{s}$, and this must be responded by $\langle s, \bar{s} \rangle \xrightarrow{a_{\langle s, \bar{s} \rangle}} \bar{s}'$. But we again have $\bar{s} \npreceq \bar{s}'$ by the "least-rank" assumption, which contradicts with $s \preceq s'$.

(b) It is easy to verify that the following set is a bisimulation in $\mathcal{L}(\mathcal{G})$:

$$I \cup \{(s, s') \mid s \in V \smallsetminus Win_\exists\} \cup \{(\langle s, X \rangle, \langle s, \bar{s} \rangle) \mid s \in V_\forall \smallsetminus Win_\exists, \bar{s} \in V \smallsetminus Win_\exists\}$$

where $I = \{(s, s) \mid s \in S\}$. $\square$

We note that the transitions corresponding to the dashed edges in Fig. 4 could be omitted if we only wanted to show that $s \in Win_\exists$ iff $s \npreceq s'$.

4.2   SOCNRG Reduces to Behavioural Relations on SOCNs

We now note that the LTS $\mathcal{L}(\mathcal{G}_\mathcal{N})$ "mimicking" the r-game $\mathcal{G}_\mathcal{N}$ associated with a socn-r-game $\mathcal{N}$ (recall (2)) can be presented as $\mathcal{L}_{\mathcal{N}'}$ for a SOCN $\mathcal{N}'$ (recall (1)) that is efficiently constructible from $\mathcal{N}$:

Lemma 4. *There is a log-space algorithm that, given a socn-r-game $\mathcal{N}$, constructs a SOCN $\mathcal{N}'$ such that the LTSs $\mathcal{L}(\mathcal{G}_{\mathcal{N}})$ and $\mathcal{L}_{\mathcal{N}'}$ are isomorphic.*

Proof. We again use Figs. 3 and 4 for illustration; now s_i are viewed as control states and the bracketed parts of edge-labels are counter-changes (in binary).

Given a socn-r-game $\mathcal{N} = (Q, Q_{\exists}, \delta, p_{win})$, we first consider the r-game $\mathcal{N}^{csg} = (Q, Q_{\exists}, \rightarrow, \{p_{win}\})$ ("the control-state game of $\mathcal{N}$") arising from $\mathcal{N}$ by *forgetting the counter-changes*; hence $q \rightarrow \bar{q}$ iff there is a rule $q \xrightarrow{z} \bar{q}$. In fact, we will assume that there is at most one rule $q \xrightarrow{z} \bar{q}$ in δ (of $\mathcal{N}$) for any pair $(q, \bar{q}) \in Q \times Q$; this can be achieved by harmless modifications.

We construct the (finite) LTS $\mathcal{L}(\mathcal{N}^{csg})$ ("mimicking" $\mathcal{N}$). Hence each $q \in Q$ has the copies q, q' in $\mathcal{L}(\mathcal{N}^{csg})$, and other states are added (as also depicted in Fig. 4 where s_i are now in the role of control states); there are also the respective labelled transitions in $\mathcal{L}(\mathcal{N}^{csg})$, with labels $a_{\langle q, \bar{q}\rangle}$, a_c, a_{win}.

It remains to add the counter changes (integer increments and decrements in binary), to create the required SOCN $\mathcal{N}'$. For $q \in Q_{\exists}$ this adding is simple, as depicted in Fig. 3: if $q \xrightarrow{z} \bar{q}$ (in $\mathcal{N}$), then we simply extend the label $a_{\langle q, \bar{q}\rangle}$ in $\mathcal{L}(\mathcal{N}^{csg})$ with z; for $q \xrightarrow{a_{\langle q, \bar{q}\rangle}} \bar{q}$ and $q' \xrightarrow{a_{\langle q, \bar{q}\rangle}} \bar{q}'$ in $\mathcal{L}(\mathcal{N}^{csg})$ we get $q \xrightarrow{a_{\langle q, \bar{q}\rangle}, z} \bar{q}$ and $q' \xrightarrow{a_{\langle q, \bar{q}\rangle}, z} \bar{q}'$ in $\mathcal{N}'$.

For $q \in Q_{\forall}$ (where $Q_{\forall} = Q \smallsetminus Q_{\exists}$) it is tempting to the same, i.e. to extend the label $a_{\langle q, \bar{q}\rangle}$ with z when $q \xrightarrow{z} \bar{q}$, and extend a_c with 0. But this might allow cheating for Defender: she could thus mimic choosing a transition $q(k) \xrightarrow{x} \bar{q}(k+x)$ even if $k + x < 0$. This is avoided by the modification that is demonstrated in Fig. 4 (by $x = x' + x''$, etc.); put simply: Defender must immediately prove that the transition she is choosing to mimic is indeed performable. Formally, if $X = \{\bar{q} \mid q \rightarrow \bar{q}\} \neq \emptyset$ (in $\mathcal{L}(\mathcal{N}^{csg})$), then in $\mathcal{N}'$ we put $q \xrightarrow{a_c, 0} \langle q, X\rangle$ and $\langle q, X\rangle \xrightarrow{a_{\langle q, \bar{q}\rangle}, z} \bar{q}$ for each $q \xrightarrow{z} \bar{q}$ (in $\mathcal{N}$); for each $q \xrightarrow{z} \bar{q}$ we also define $z' = \min\{z, 0\}$, $z'' = \max\{z, 0\}$ and put $q' \xrightarrow{a_c, z'} \langle q, \bar{q}\rangle$, $\langle q, \bar{q}\rangle \xrightarrow{a_{\langle q, \bar{q}\rangle}, z''} \bar{q}'$. Then for any pair $q \xrightarrow{z} \bar{q}$, $q \xrightarrow{\bar{z}} \bar{\bar{q}}$ where $\bar{q} \neq \bar{\bar{q}}$ we put $\langle q, \bar{q}\rangle \xrightarrow{a_{\langle q, \bar{q}\rangle}, \bar{z} - z'} \bar{\bar{q}}$.

Finally, $p_{win} \xrightarrow{a_{win}} p_{win}$ in $\mathcal{L}(\mathcal{N}^{csg})$ is extended to $p_{win} \xrightarrow{a_{win}, 0} p_{win}$ in $\mathcal{N}'$. $\square$

Recalling the EXPSPACE-hardness of SOCNRG (from [11] or from Theorem 1), Lemmas 3 and 4 yield:

Theorem 5. *For succinct labelled one-counter nets (SOCNs), deciding any relation containing bisimulation equivalence and contained in simulation preorder is EXPSPACE-hard.*

5   Additional Remarks

Theorem 5 shows the hardness, but simulation preorder on succinct one-counter nets and bisimulation equivalence (even) on succinct one-counter automata are also in EXPSPACE (as follows by the membership of their unary versions in PSPACE), hence they are EXPSPACE-complete. Hunter [11] also shows that various extensions of the problem SOCNRG are in EXPSPACE as well.

We recall that our EXPSPACE-hardness proof of ECG (in Sect. 3) is, in fact, a particular instance of a simple general method. A slight modification also yields a proof of EXPTIME-hardness of countdown games that is an alternative to the proof in [17].

One particular application of countdown games was shown by Kiefer [18] who modified them to show EXPTIME-hardness of bisimilarity on BPA processes. Our EXPSPACE-complete modification does not seem easily implementable by BPA processes, hence the EXPTIME-hardness result in [18] has not been improved here. (The known upper bound for bisimilarity on BPA is 2-EXPTIME.)

We have not discussed the upper bounds here. The proofs in [1,10,14] reveal a periodicity of simulation preorder, captured by "linear-belt" theorems. It is worth to note that the period of a simulation-belt can be double-exponential in the size of the respective succinct one-counter net. This is derivable by recalling the reduction from exponential-space Turing machines and by noting that we can choose a machine M such that for every n there is an accepted word of length n for which M performs $2^{2^{\Omega(n)}}$ steps.

Finally we mention that the involved result in [6] shows that, given any fixed language L in EXPSPACE, for any word w (in the alphabet of L) we can construct a succinct one-counter automaton that performs a computation which is accepting iff $w \in L$. Such a computation needs to access concrete bits in the (reversed) binary presentation of the counter value. A straightforward direct access to such bits is destructive (the counter value is lost after the bit is read) but this can be avoided: instead of a "destructive reading" the computation just "guesses" the respective bits, and it is forced to guess correctly by a carefully constructed CTL formula that is required to be satisfied by the computation. This result is surely deeper than the EXPSPACE-hardness of existential countdown games, though the former does not seem to entail the latter immediately.

References

1. Abdulla, P.A., Čerāns, K.: Simulation is decidable for one-counter nets. In: Sangiorgi, D., de Simone, R. (eds.) CONCUR 1998. LNCS, vol. 1466, pp. 253–268. Springer, Heidelberg (1998). https://doi.org/10.1007/BFb0055627
2. Böhm, S., Göller, S., Jančar, P.: Equivalence of deterministic one-counter automata is NL-complete. In: STOC 2013, pp. 131–140. ACM (2013)
3. Böhm, S., Göller, S., Jančar, P.: Bisimulation equivalence and regularity for real-time one-counter automata. J. Comput. Syst. Sci. 80(4), 720–743 (2014). Preliminary versions appeared at CONCUR 2010 and MFCS 2011

4. Chandra, A.K., Kozen, D., Stockmeyer, L.J.: Alternation. J. ACM **28**(1), 114–133 (1981)
5. Demri, S., Lazić, R., Sangnier, A.: Model checking freeze LTL over one-counter automata. In: Amadio, R. (ed.) FoSSaCS 2008. LNCS, vol. 4962, pp. 490–504. Springer, Heidelberg (2008). https://doi.org/10.1007/978-3-540-78499-9_34
6. Göller, S., Haase, C., Ouaknine, J., Worrell, J.: Model checking succinct and parametric one-counter automata. In: Abramsky, S., Gavoille, C., Kirchner, C., Meyer auf der Heide, F., Spirakis, P.G. (eds.) ICALP 2010 Part II. LNCS, vol. 6199, pp. 575–586. Springer, Heidelberg (2010). https://doi.org/10.1007/978-3-642-14162-1_48
7. Göller, S., Lohrey, M.: Branching-time model checking of one-counter processes. In: STACS 2010. LIPIcs, vol. 5, pp. 405–416. Schloss Dagstuhl - Leibniz-Zentrum fuer Informatik (2010)
8. Göller, S., Mayr, R., To, A.W.: On the computational complexity of verifying one-counter processes. In: LICS 2009, pp. 235–244. IEEE Computer Society (2009)
9. Haase, C., Kreutzer, S., Ouaknine, J., Worrell, J.: Reachability in succinct and parametric one-counter automata. In: Bravetti, M., Zavattaro, G. (eds.) CONCUR 2009. LNCS, vol. 5710, pp. 369–383. Springer, Heidelberg (2009). https://doi.org/10.1007/978-3-642-04081-8_25
10. Hofman, P., Lasota, S., Mayr, R., Totzke, P.: Simulation problems over one-counter nets. Log. Methods Comput. Sci. **12**(1), 46 pp. (2016). Preliminary versions appeared at FSTTCS 2013, LICS 2013, and in Totzke's Ph.D. thesis (2014)
11. Hunter, P.: Reachability in succinct one-counter games. In: Bojańczyk, M., Lasota, S., Potapov, I. (eds.) RP 2015. LNCS, vol. 9328, pp. 37–49. Springer, Cham (2015). https://doi.org/10.1007/978-3-319-24537-9_5
12. Jančar, P., Kučera, A., Moller, F.: Simulation and bisimulation over one-counter processes. In: Reichel, H., Tison, S. (eds.) STACS 2000. LNCS, vol. 1770, pp. 334–345. Springer, Heidelberg (2000). https://doi.org/10.1007/3-540-46541-3_28
13. Jančar, P., Kučera, A., Moller, F., Sawa, Z.: DP lower bounds for equivalence-checking and model-checking of one-counter automata. Inf. Comput. **188**(1), 1–19 (2004)
14. Jančar, P., Moller, F., Sawa, Z.: Simulation problems for one-counter machine. In: Pavelka, J., Tel, G., Bartošek, M. (eds.) SOFSEM 1999. LNCS, vol. 1725, pp. 404–413. Springer, Heidelberg (1999). https://doi.org/10.1007/3-540-47849-3_28
15. Jančar, P., Sawa, Z.: A note on emptiness for alternating finite automata with a one-letter alphabet. Inf. Process. Lett. **104**(5), 164–167 (2007)
16. Jančar, P., Srba, J.: Undecidability of bisimilarity by defender's forcing. J. ACM **55**(1), 5:1–5:26 (2008)
17. Jurdzinski, M., Sproston, J., Laroussinie, F.: Model checking probabilistic timed automata with one or two clocks. Log. Methods Comput. Sci. **4**(3), 28 pp. (2008)
18. Kiefer, S.: BPA bisimilarity is EXPTIME-hard. Inf. Process. Lett. **113**(4), 101–106 (2013)
19. Kučera, A.: Efficient verification algorithms for one-counter processes. In: Montanari, U., Rolim, J.D.P., Welzl, E. (eds.) ICALP 2000. LNCS, vol. 1853, pp. 317–328. Springer, Heidelberg (2000). https://doi.org/10.1007/3-540-45022-X_28
20. Mayr, R.: Undecidability of weak bisimulation equivalence for 1-counter processes. In: Baeten, J.C.M., Lenstra, J.K., Parrow, J., Woeginger, G.J. (eds.) ICALP 2003. LNCS, vol. 2719, pp. 570–583. Springer, Heidelberg (2003). https://doi.org/10.1007/3-540-45061-0_46
21. Reichert, J.: On the complexity of counter reachability games. Fundam. Inform. **143**(3–4), 415–436 (2016)

22. Serre, O.: Parity games played on transition graphs of one-counter processes. In: Aceto, L., Ingólfsdóttir, A. (eds.) FoSSaCS 2006. LNCS, vol. 3921, pp. 337–351. Springer, Heidelberg (2006). https://doi.org/10.1007/11690634_23
23. Srba, J.: Beyond language equivalence on visibly pushdown automata. Log. Methods Comput. Sci. 5(1), 22 pp. (2009)
24. Valiant, L.G., Paterson, M.: Deterministic one-counter automata. J. Comput. Syst. Sci. 10(3), 340–350 (1975)

Revisiting MU-Puzzle. A Case Study in Finite Countermodels Verification

Alexei Lisitsa[(✉)]

Department of Computer Science, University of Liverpool, Liverpool, UK
A.Lisitsa@liverpool.ac.uk

Abstract. In this paper we consider well-known MU puzzle from *Goedel, Escher, Bach: An Eternal Golden Braid* book by D. Hofstadter, as an infinite state safety verification problem for string rewriting systems. We demonstrate fully automated solution using finite countermodels method (FCM). We highlight advantages of FCM method and compare it with alternatives methods using regular invariants.

1 MIU System and MU Puzzle

In his famous book *Goedel, Escher, Bach: An eternal Golden Braid, 1979*, Douglas Hofstadter introduced a simple formal system, named MIU-system, which operates on strings made of three symbols, M, I and U. The system consists of one axiom, that is MI and four derivation rules:

I. If xI is a theorem, so is xIU.
II. If Mx is theorem, so is Mxx.
III. In any theorem III can be replaced by U.
IV. UU can be dropped from any theorem.

In other words, MIU system is a *string rewriting system* with an initial string MI and the set of rewriting rules $R = \{xI \Rightarrow xIU; Mx \Rightarrow Mxx; xIIIy \Rightarrow xUy; xUUy \Rightarrow xy\}$. We denote the language generated by this rewriting system by L_{MIU}. From now on we use interchangeably expressions "a string S is a theorem of MIU system" and "string S belongs to the language L_{MIU}".

MU puzzle is a specific problem about MIU system, that is "Is MU a theorem of MIU system?" The problem is discussed at length in [4] and the answer is negative. It follows from a simple necessary condition: "the number of I symbols in any string in L_{MIU} cannot be multiple of three". The authors of [11] show that this condition augmented with structural requirement that any MIU theorem should start with M followed by an arbitrary word in I's and U's is also sufficient, obtaining thereby a simple decision procedure for MIU theorems.[1]

We show here an alternative way to get an answer (with a proof) for MU puzzle automatically, from first principles and not assuming the knowledge of the

---

[1] They also notice that Hofstadter was aware about the decision procedure, but never formally wrote a proof.

© Springer Nature Switzerland AG 2018
I. Potapov and P.-A. Reynier (Eds.): RP 2018, LNCS 11123, pp. 75–86, 2018.
https://doi.org/10.1007/978-3-030-00250-3_6

decision procedure. First notice that there are infinitely many theorems in MIU, so the negative answer can not be obtained just by exhaustion of all derivable strings. It is essentially infinite state verification problem.

2   First-Order Logic Encoding and Disproving for MIU

In order to deal with a problem automatically we formulate a natural theory T_{MIU} in first-order logic which encodes the rewriting process. The vocabulary of the T_{MIU} consists of one unary predicate symbol T binary functional symbol $*$ which we use in infix notation an three constants M, I and U. Intended meaning of T(x) is "x is a theorem of MIU" and $*$ denotes concatenation to be used to build strings out of constants.

The theory T_{MIU} consist the following axioms:

1. $(x * y) * z = x * (y * z)$ (associativity of concatenation);
2. $e * x = x$;
3. $x * e = x$;
4. $T(M * I)$ (MI is a theorem of MIU);
5. $T(x * I) \rightarrow T(x * I * U)$ (rule I of MIU);
6. $T(M * x) \rightarrow T(M * x * x)$ (rule II of MIU);
7. $T(x * I * I * I * y) \rightarrow T(x * U * y)$ (rule III of MIU);
8. $T(x * U * U * y) \rightarrow T(x * y)$ (rule IV of MIU).

Now we have a simple proposition.

Proposition 1. *If $S \in L_{MIU}$ then $T_{MIU} \vdash_{FO} T(t_S)$ where t_S is a term encoding of S; e.g. $t_{IUM} \equiv I * U * M$.*

Proof. Straightforward induction on the derivation of S in MIU. Indeed $T(M * I) \equiv T(t_{MI})$ is an axiom of T_{MIU}, so the base of induction holds true: $T_{MIU} \vdash_{FO} T(t_{MI})$. Assume the proposition holds true for a string S in L_{MIU}, and S' is obtained from S by application of the rule I. Then we have: (1) $T_{MIU} \vdash T(t_S)$ by induction assumption; (2) $T_{MIU} \vdash_{FO} T(t_S) \rightarrow T(t_{S'})$ by axiom 3 and finally, (3) $T_{MIU} \vdash T(t_{S'}))$ by Modus Ponens applied to (2) and (3). The cases of S' obtained from S by rules II–IV are considered similarly using axioms 4–6. The step of induction is proven.

We have an immediate corollary.

Corollary 1.   – *If $T(t_S)$ is not FO provable from T_{MIU}, that is $T_{MIU} \nvdash_{FO} T(t_S)$ then $S \notin L_{MIU}$;*
– *For any non-ground term $t(\bar{x})$ in vocabulary $\{*, M, I, U\}$ over the set of variables X, if $T_{MIU} \nvdash_{FO} \exists \bar{x} T(t(\bar{x}))$ then none of S such that t_S is a ground instance of $t(\bar{x})$ belongs to L_{MIU}.*

Returning to MU puzzle it should be clear now that to answer its question negatively it is sufficient to find a countermodel for $T_{MIU} \rightarrow T(t_{MU})$, or, in other words, a model for $T_{MIU} \land \neg T(t_{MU})$. We delegate this problem to Mace4

[9], the automated finite model finder for first-order logic. The countermodel of size 3 is found in 0.05s[2]. The property is proven: MU is not a theorem of MIU system. On the face of it, we have a simple logical argument: should MU be a theorem of MIU the formula $T(t_{MU})$ would be provable from T_{MIU}; since we found a countermodel for $T_{MIU} \rightarrow T(t_{MU})$, this is impossible. This argument does not explain though "the reasons" for impossibility. To recover more detailed argument let us have a look at the generated countermodel.

The domain $\mathcal{M}$ of the model is the set $0, 1, 2$ the interpretations of constants M, I and U are $0, 0$ and 1, respectively. The interpretation $[*]$ of concatenation (monoid) operation $*$ is given by the table.

```
[*]   0  1 2
      ------
  0  |2,0,1
  1  |0,1,2
  2  |1,2,0
```

The interpretation $[T]$ of unary predicate T includes elements $1, 2$ of the domain, meaning T is true on $1, 2$ and false on 0. Now we notice that the model provides with an interpretation $[t_S] \in \{0, 1, 2\}$ of any term t_S. The following property holds: for any theorem S of MIU the interpretation $[t_S]$ should be an element of $\{1, 2\} = [T]$ (as $\mathcal{M}$ is a model of T_{MIU} and by Proposition 1). Returning to MU puzzle, we have interpretation $[t_{MU}] = [M * U] = 0[*]1 = 0 \notin \{1, 2\} = [T]$. Therefore MU is not a theorem of MIU. In summary, the interpretation $[*]$ above defines the set of strings $L_{\mathcal{M}} = \{s \mid [t_s]_{\mathcal{M}} \in \{0, 1\}\}$ for which (1) $L_{MIU} \subseteq L_{\mathcal{M}}$; (2) $MU \notin L_{\mathcal{M}}$. Thus, $L_{\mathcal{M}}$ is an invariant separating the theorems of MIU system and the string in question, MU. It is easy to see also that the invariant is a *regular* language. It follows from an algebraic characterization of the regular languages by inverse homomorphisms of finite monoids, see e.g. [6].

Interestingly, $L_{\mathcal{M}} \neq L_{MIU}$ as, for example, $[M * M] = 2 \in [T]$ hence $MM \in L_{\mathcal{M}}$ but $MM \notin L_{MIU}$ by decision procedure of [11]. Applying our method to show $MM \notin L_{MIU}$ we formulate the formula to disprove: $T_{MIU} \rightarrow T(M * M)$. Mace4 finds a countermodel $L_{\mathcal{M}'}$ of size 2, with the domain $\{0, 1\}$, the interpretations of constants M, I and U as $1, 0$ and 0, respectively; the interpretation $[T]$ of $T = \{1\}$. the interpretation of $*$ is given by the table.

```
[*]    0  1
      ----
  0  |0,1
  1  |1,0
```

The corresponding invariant $\{s \mid [t_s]_{\mathcal{M}'} = 1\}$ captures the "oddness" of M count in strings, which is sufficient to separate MM from L_{MIU}.

---

[2] We used Prover9 and Mace4 version 0.5 (December 2007) [9] running on AMD A6-3410MX APU 1.60Ghz, RAM 4 GB, Windows 7 Enterprise.

2.1   Assumptions on Model Building Procedure

Before we continue with the exploration of the MIU problem we need to make some implicit so far assumptions on Mace4 procedure explicit. In what follows we assume that Mace4 is used with the default iterative search strategy: the search for a model starts with the size 2; if no model is found by an exhaustive search of models of a certain size, the size is increased by 1 and the search continues. It is clear then if a finite model is found it is a *minimal* one in a partial order defined as $\mathcal{M}_1 \leq \mathcal{M}_2$ iff $|M_1| \leq |M_2|$, where M_i denotes a domain of $\mathcal{M}_i$. Another obvious assumption is that the implementation of Mace4 is correct, and in particular if it returns something, it is indeed a minimal model. We notice that for any concrete result produced by Mace4, its verification is a finite problem which can be tackled independently. The checking that the result is indeed a required model can be done in polynomial in the size of the model time, while checking minimality requires an exponential in the size of the model time. In what follows we assume correctness of Mace4 procedure.

2.2   Exact Invariant by Model Building

The natural question appears as to whether by an appropriate choice of target "non-theorems" of MIU one can get a minimal countermodel defining an exact invariant coinciding with L_{MIU}. We answer this question positively by introducing "disjunctive targets" formulas.

Consider the formula

$$\varphi_d \equiv \exists x \exists y T(M * x * M * y) \vee \exists x T(I * x) \vee \exists x T(U * x) \vee T(M * U)$$

Notice that neither MU (occurring in the last disjunct of the formula) nor any of the ground instances of existential disjuncts are elements of L_{MIU} (by decision procedure). For the formula $T_{MIU} \rightarrow \varphi_d$ finite model finder Mace4 finds a minimal countermodel $\mathcal{M}''$ of size 8.

The domain of $\mathcal{M}''$ is the set $\{0, 1, 2, 3, 4, 5, 6, 7\}$; the interpretations of the constants M, I, U and e are $1, 0, 2$ and 3 respectively. The interpretation $[T]$ of T is $\{4, 5\}$ and $[*]$ is given by the following multiplication table.

```
          0 1 2 3 4 5 6 7
[*]       ------------------              .
    0 |  6,7,0,0,7,7,2,7,
    1 |  4,7,1,1,7,7,5,7,
    2 |  0,7,2,2,7,7,6,7,
    3 |  0,1,2,3,4,5,6,7,
    4 |  5,7,4,4,7,7,1,7,
    5 |  1,7,5,5,7,7,4,7,
    6 |  2,7,6,6,7,7,0,7,
    7 |  7,7,7,7,7,7,7,7
```

Proposition 2. *The invariant $L_{\mathcal{M}''}$ defined by the countermodel $\mathcal{M}''$ coincides with L_{MIU}, that is the interpretation of any term t_S belongs to the interpretation $[T]$ of T iff "S starts with symbol M, followed by an arbitrary word in symbols I and U with a number of I being not multiple of 3."*

Proof. By the decision procedure of [11] any non-theorem of MIU system is either (i) a word starting with I letter; or (ii) a word starting with U letter; or (iii) a word starting from M letter and having two or more M letters; or (iv) a word starting from M letter following by a word in I and U letters with multiplicity of I being multiple of 3. All other words are theorems. It follows that to prove Proposition, it is sufficient to show that for a term encoding τ of any word in (i)–(iv) $T(\tau)$ is false in $\mathcal{M}''$. First notice that for any ground instance τ of $M*x*M*y$, $I*x$, or $U*x$, that is term encoding of any word in (i)–(iii) the formula $T(\tau)$ is false in $\mathcal{M}''$ by $\mathcal{M}'' \not\models \varphi_d$. We notice also that the same argument can be applied for a word MU from category (iv). What about all other words from the category (iv)? We have

Lemma 1. *For a term encoding τ of any word in (iv) $\mathcal{M}'' \not\models T(\tau)$.*

The Lemma can be proved by a straightforward induction on the length of the word. We demonstrate an alternative argument using automated reasoning. Consider a theory C_{IV} ("Condition IV") consisting of the following formulae.

- $C(e)$
- $C(x) \rightarrow C(x * I * I * I)$
- $C(x) \rightarrow C(x * U)$
- $C(x * y) \rightarrow C(y * x)$

If a word belongs to a category (iv), that is it is a word starting from M letter following by a word in I and U letters with multiplicity of I being multiple of 3 then for its term encoding τ we have $C_{IV} \vdash C(\tau)$. Now we update target formula φ_d to

$$\varphi_d^{C_{IV}} \equiv \exists x \exists y T(M*x*M*y) \vee \exists x T(I*x) \vee \exists x T(U*x) \vee \exists x (T(M*x) \wedge C(x))$$

and consider disproving task for $T_{MIU} \wedge C_{IV} \rightarrow \varphi_d^{C_{IV}}$. The model finder returns a countermodel which we denote $\mathcal{M}''_C$ which in fact is an extension of $\mathcal{M}''$ above. That means it has the same domain as $\mathcal{M}''$ and the same interpretations of M,U,I, $*$ and e. The only difference is an additional interpretation of C which is $[C] = \{2,3\}$. To conclude the proofs of Lemma, and Proposition we notice for any word w from category (iv) and its term encoding τ_w we have $\mathcal{M}''_C \not\models T(\tau_w)$ (by $\mathcal{M}''_C \not\models \exists x (T(M*x) \wedge C(x))$) and $\mathcal{M}'' \models T(\tau) \Leftrightarrow \mathcal{M}''_C \models T(\tau)$ for any ground τ.   □

By classical Herbrand theorem we have the following

Corollary 2. *There is a finite set of words $w, \dots w_n$ such that L_{MIU} coincides with an invariant defined by a countermodel for $T_{MIU} \rightarrow \vee_{i=1\dots n} T(\tau_{w_i})$.*

The natural question appears as to whether one can simplify the definition of L_{LIU} via minimal countermodels further and obtain it using a single target formula $T(\tau)$ with a ground term τ.

Proposition 3. *There is no single target formula $T(\tau)$ with a ground τ for which a minimal countermodel defines L_{MIU}.*

Proof. By the decision procedure of [11] any non-theorem of MIU system is either (i) a word starting with I letter; or (ii) a word starting with U letter; or (iii) a word starting from M letter and having two or more M letters; or (iv) a word starting from M letter following by a word in I and U letters with multiplicity of I being multiple of 3. We consider all these cases in their turn.

(i) For the formula $T_{MIU} \rightarrow \exists x T(I * x)$ Mace4 model finder generates the following minimal countermodel (in Mace4 output syntax):

```
interpretation( 2, [number = 1,seconds = 0], [
    function(*(_,_), [
        0,0,
        1,1]),
    function(I, [0]),
    function(M, [1]),
    function(U, [0]),
    relation(T(_), [0,1])]).
```

It follows that for any ground instance τ of $I * x$ the above is a countermodel, and therefore the minimal countermodel for any such τ is no larger than the above model.

(ii) For the formula $T_{MIU} \rightarrow \exists x T(U * x)$ Mace4 model finder generates the same minimal countermodel as presented above in (i). The same argument follows.

(iii) For the formula $T_{MIU} \rightarrow \exists x \exists y R(M * x * M * y)$ Mace4 generates the following minimal countermodel.

```
interpretation( 3, [number = 1,seconds = 0], [
    function(*(_,_), [
        0,1,2,
        1,2,2,
        2,2,2]),
    function(I, [0]),
    function(M, [1]),
    function(U, [0]),
    relation(T(_), [0,1,0])]).
```

It follows that for any ground instance τ of $M * x * M * y$ the above is a countermodel, and therefore the minimal countermodel for any such τ is no larger than the above model.

(iv) For the formula $T_{MIU} \wedge C_{IV} \rightarrow \exists x T(M * x) \wedge C(x))$ Mace4 model finder generates the following countermodel $\mathcal{M}_C$.

```
interpretation( 3, [number = 1,seconds = 0], [
    function(*(_,_), [
        2,0,1,
        0,1,2,
        1,2,0]),'
    function(I, [0]),
    function(M, [0]),
    function(U, [1]),
    function(e, [1]),
    relation(C(_), [0,1,0]),
    relation(T(_), [0,1,1])]).
```

It follows that for any word w from (iv) the minimal countermodel for $T_{MIU} \rightarrow T(\tau_w)$ is not larger than $\mathcal{M}$ obtained from $\mathcal{M}_C$ above by omitting the interpretation of C.   □

We can strengthen Proposition 3.

Proposition 4. *There are no two words w_1 and w_2 such that a minimal countermodel for $T_{MIU} \rightarrow \vee_{i=1,2}T(\tau_{w_i})$ defines L_{MIU}.*

Proof. By the argument used in the proof of Proposition 3, it is sufficient to show that a minimal countermodel for $T_{MIU} \wedge C_{IV} \rightarrow \phi'_d$ where ϕ'_d is any two disjunct subformula of $\varphi_d^{C_{IV}}$, is less (in the partial order $\leq$) than a minimal model for $T_{MIU} \wedge C_{IV} \rightarrow \varphi_d^{C_{IV}}$. This is indeed the case. The minimal countermodels for all such subformulae found by Mace4 are shown in the Appendix.   □

2.3 Variations: Symmetric MIU Problem

Let MIU^{SYM} be a symmetric variant of MIU, that is a string rewriting system where all rules of MIU can be applied in forward and backward direction. The first-order theory $T_{MIU^{SYM}}$ is naturally defined as

1. $(x * y) * z = x * (y * z)$ (associativity of concatenation);
2. $e * x = x$;
3. $x * e = x$;
4. $T(M * I)$ (MI is a theorem of MIU^{SYM});
5. $T(x * I) \leftrightarrow T(x * I * U)$ (rule I^{SYM});
6. $T(M * x) \leftrightarrow T(M * x * x)$ (rule II^{SYM});
7. $T(x * I * I * I * y) \leftrightarrow T(x * U * y)$ (rule III^{SYM})
8. $T(x * U * U * y) \leftrightarrow T(x * y)$ (rule IV^{SYM}).

The next proposition strengthens the answer to the original MU puzzle.

Proposition 5. $MU \notin L_{MIU^{SYM}}$ and $L_{MIU} \neq L_{MIU^{SYM}}$.

Proof. (i) Consider the formula $T_{MIU^{SYM}} \rightarrow T(M * U)$. Mace4 finds the following countermodel for it.

```
interpretation( 3, [number = 1,seconds = 0], [
    function(*(_,_), [
        2,0,1,
        0,1,2,
        1,2,0]),
    function(I, [0]),
    function(M, [0]),
    function(U, [1]),
    function(e, [1]),
    relation(T(_), [0,1,1])]).
```

By the analogue of Corollary 1 for symmetric rewriting it follows that $MU \notin L_{MIU^{SYM}}$.

(ii) $UUMI \in L_{MIU^{SYM}}$ (apply IV^{SYM} to the initial word MI), while $UUMI \notin L_{MIU}$. □

3   String Rewriting and Regular Invariants

The solution of MU puzzle we presented here is an instance of the application of very general finite countermodel safety verification method (FCM) from [3,5,7,8, 10]. Its instantiation to the general string rewriting systems is discussed in [7].

Abstracting from the details of particular model finding procedures the work of FCM method can be ascribed as follows. Given a string rewriting system S and a target string s, find a (minimal) regular separator, that is a regular set R subsuming all reachable strings R_S and disjoint with a target string s, that is $R_S \subseteq R \wedge s \notin R$. A few observations are in order.

Obviously, if the set of all reachable strings R_S is regular then R_S itself is a regular separator for any target non-reachable string s. For a target s though such R_S is not necessarily minimal[3] separator. Recall MM can be separated from L_{MIU} by a very simple, as compared with the whole regular L_{MIU}, *oddness/parity* separator. This is the one of the distinctive features of FCM as compared with other safety verification methods using regular invariants, notably regular model checking [1], that it does not rely on computing regular approximations of *all reachable states* independently of safety conditions. It rather takes an account of a safety condition (target non-reachable strings in our case) and searches for a separator. That may lead to favorable performance as was noticed in [8]. We showed in this paper that in particular case of MU puzzle, by careful choice of target formulae/(sets of) strings one can use FCM to find the set of all reachable strings L_{MIU}. In this process however we used previous knowledge of L_{MIU} and applied ad hoc reasoning.

Question 1. Is it possible to have systematic procedure using first-order proving/disproving for generation of regular approximations for sets of reachable strings for string rewriting systems?

---

[3] In a reasonably defined partial order. Instead of a partial order $\leq$ motivated by the iterative finite model building procedure, one may consider a partial order defined by inclusion of corresponding languages.

We showed also that for MIU system the set of reachable strings L_{MIU} can not be generated by FCM using less than three target strings. For a string rewriting system S with a regular R_S, one can define a *separation dimension* $Dim(S)$ as a minimal number of target words required to generate R_S. In these terms $Dim(MIU) \geq 3$. We leave the following questions for further exploration elsewhere.

Question 2. What is Dim(MIU)?

Question 3. Does $Dim(S)$ form proper hierarchy, that is if there is an infinite sequence $S_i, i \geq 1$ of string rewriting systems, such that $Dim(S_i) = i$?

4   Conclusion

We have shown in this paper how to solve MU puzzle by first-order theorem disproving (finite model finding) fully automatically and from the first principles. As far we are aware, no fully automatic solution of this puzzle has been presented in the literature so far. We have further shown that the known decision procedure can be re-interpreted in terms of a single finite countermodel. MU puzzle in a instance of an infinite state verification problem and as such it was used as a case study to illustrate the verification methods based on Counter Example Guided Refinement (CEGAR) in [2]. The verification presented in [2] was not fully automated and required a creative step in the choice of invariants. The efficiency of FCM method can be explained in part by targeted building of regular separators for particular targets, instead of regular approximations of all reachable states. Further investigation of related questions in terms of separation dimension is a topic for future work.

Appendix

To the Proof of Proposition 4. We present here all minimal countermodels found by Mace4 for all formulae $T_{MIU} \wedge C_{IV} \rightarrow \phi'_d$ with ϕ'_d being two disjunct subformulae of $\varphi_d^{C_{IV}} \equiv \exists x \exists y T(M * x * M * y) \vee \exists x T(I * x) \vee \exists x T(U * x) \vee \exists x (T(M * x) \wedge C(x))$. Notice that all of them are less wrt $\leq$ than a minimal countermodel for $T_{MIU} \wedge C_{IV} \rightarrow \varphi_d^{C_{IV}}$ whose domain size is 8.

(1) $\phi'_d \equiv \exists x \exists y T(M * x * M * y) \vee \exists x T(I * x)$

```
interpretation( 4, [number = 1,seconds = 0], [
    function(*(_,_), [
        0,3,0,3,
        1,3,1,3,
        0,1,2,3,
        3,3,3,3]),
    function(I, [0]),
    function(M, [1]),
```

```
function(U, [0]),
function(e, [2]),
relation(C(_), [1,0,1,0]),
relation(R(_), [0,1,0,0])]).
```

(2) $\phi'_d \equiv \exists x \exists y T(M * x * M * y) \vee \exists x T(U * x)$

```
interpretation( 4, [number = 1,seconds = 0], [
    function(*(_,_), [
        0,3,0,3,
        1,3,1,3,
        0,1,2,3,
        3,3,3,3]),
    function(I, [0]),
    function(M, [1]),
    function(U, [0]),
    function(e, [2]),
    relation(C(_), [1,0,1,0]),
    relation(R(_), [0,1,0,0])]).
```

(3) $\phi'_d \equiv \exists x \exists y T(M * x * M * y) \vee \exists x (T(M * x) \wedge C(x))$

```
interpretation( 7, [number = 1,seconds = 0], [
    function(*(_,_), [
        5,1,0,3,4,2,6,
        3,4,1,4,4,6,4,
        0,1,2,3,4,5,6,
        6,4,3,4,4,1,4,
        4,4,4,4,4,4,4,
        2,1,5,3,4,0,6,
        1,4,6,4,4,3,4]),
    function(I, [0]),
    function(M, [1]),
    function(U, [2]),
    function(e, [2]),
    relation(C(_), [0,0,1,0,0,0,0]),
    relation(R(_), [0,0,0,1,0,0,1])]).
```

(4) $\phi'_d \equiv \exists x T(I * x) \vee \exists x T(U * x)$

```
interpretation( 3, [number = 1,seconds = 0], [
    function(*(_,_), [
        0,0,0,
        1,1,1,
        0,1,2]),
    function(I, [0]),
    function(M, [1]),
```

```
      function(U, [0]),
      function(e, [2]),
      relation(C(_), [1,1,1]),
      relation(R(_), [0,1,0])]).
```

(5) $\phi'_d \equiv \exists x T(I * x) \vee \exists x (T(M * x) \wedge C(x))$

```
interpretation( 7, [number = 1,seconds = 0], [
      function(*(_,_), [
         4,0,0,4,6,6,0,
         3,1,1,3,5,5,1,
         0,1,2,3,4,5,6,
         5,3,3,5,1,1,3,
         6,4,4,6,0,0,4,
         1,5,5,1,3,3,5,
         0,6,6,0,4,4,6]),
      function(I, [0]),
      function(M, [1]),
      function(U, [2]),
      function(e, [2]),
      relation(C(_), [0,1,1,0,0,0,1]),
      relation(R(_), [0,0,0,1,0,1,0])]).
```

(6) $\phi'_d \equiv \exists x T(U * x) \vee \exists x (T(M * x) \wedge C(x))$

```
interpretation( 7, [number = 1,seconds = 0], [
      function(*(_,_), [
         5,0,0,0,5,3,3,
         4,1,1,1,4,6,6,
         0,1,2,3,4,5,6,
         0,3,3,3,0,5,5,
         6,4,4,4,6,1,1,
         3,5,5,5,3,0,0,
         1,6,6,6,1,4,4]),
      function(I, [0]),
      function(M, [1]),
      function(U, [3]),
      function(e, [2]),
      relation(C(_), [0,1,1,1,0,0,0]),
      relation(R(_), [0,0,0,0,1,0,1])]).
```

References

1. Abdulla, P.A., Jonsson, B., Nilsson, M., Saksena, M.: A survey of regular model checking. In: Gardner, P., Yoshida, N. (eds.) CONCUR 2004. LNCS, vol. 3170, pp. 35–48. Springer, Heidelberg (2004). https://doi.org/10.1007/978-3-540-28644-8_3
2. Clarke, E.M., et al.: Abstraction and counterexample-guided refinement in model checking of hybrid systems. Int. J. Found. Comput. Sci. 14(4), 583–604 (2003)
3. Goubault-Larrecq, J.: Finite models for formal security proofs. J. Comput. Secur. 18(6), 1247–1299 (2010)
4. Hofstadter, D.R.: Godel, Escher, Bach: An Eternal Golden Braid. Basic Books Inc., New York (1979)
5. Jürjens, J., Weber, T.: Finite models in FOL-based crypto-protocol verification. In: Degano, P., Viganò, L. (eds.) ARSPA-WITS 2009. LNCS, vol. 5511, pp. 155–172. Springer, Heidelberg (2009). https://doi.org/10.1007/978-3-642-03459-6_11
6. Lallement, G.: Semigroups and Combinatorial Applications. Wiley, Hoboken (1979)
7. Lisitsa, A.: Finite models vs tree automata in safety verification. In: 23rd International Conference on Rewriting Techniques and Applications, RTA 2012, Nagoya, Japan, pp. 225–239, 28 May–2 June 2012
8. Lisitsa, A.: Finite reasons for safety - parameterized verification by finite model finding. J. Autom. Reason. 51(4), 431–451 (2013)
9. McCune, W.: Prover9 and Mace4 (2005–2010). http://www.cs.unm.edu/~mccune/prover9/
10. Selinger, P.: Models for an adversary-centric protocol logic. Electron. Notes Theor. Comput. Sci. 55(1), 69–84 (2003). LACPV 2001, Logical Aspects of Cryptographic Protocol Verification (in connection with CAV 2001)
11. Swanson, L., McEliece, R.J.: A simple decision procedure for Hofstadter's MIU-system. Math. Intell. 10(2), 48–49 (1988)

Knapsack in Hyperbolic Groups

Markus Lohrey[✉]

Universität Siegen, Siegen, Germany
lohrey@eti.uni-siegen.de

Abstract. Recently knapsack problems have been generalized from the integers to arbitrary finitely generated groups. The knapsack problem for a finitely generated group G is the following decision problem: given a tuple $(g, g_1, \ldots, g_k)$ of elements of G, are there natural numbers $n_1, \ldots, n_k \in \mathbb{N}$ such that $g = g_1^{n_1} \cdots g_k^{n_k}$ holds in G? Myasnikov, Nikolaev, and Ushakov proved that for every hyperbolic group, the knapsack problem can be solved in polynomial time. In this paper, it is shown that for every hyperbolic group G, the knapsack problem belongs to the complexity class LogCFL, and it is LogCFL-complete if G contains a free group of rank two. Moreover, it is shown that for every hyperbolic group G and every tuple $(g, g_1, \ldots, g_k)$ of elements of G the set of all $(n_1, \ldots, n_k) \in \mathbb{N}^k$ such that $g = g_1^{n_1} \cdots g_k^{n_k}$ in G is effectively semilinear.

1 Introduction

In [22], Myasnikov, Nikolaev, and Ushakov initiated the investigation of discrete optimization problems, which are usually formulated over the integers, for arbitrary (possibly non-commutative) groups. One of these problems is the *knapsack problem* for a finitely generated group G: The input is a sequence of group elements $g_1, \ldots, g_k, g \in G$ (specified by finite words over the generators of G) and it is asked whether there exists a tuple $(n_1, \ldots, n_k) \in \mathbb{N}^k$ such that $g_1^{n_1} \cdots g_k^{n_k} = g$ in G. For the particular case $G = \mathbb{Z}$ (where the additive notation $n_1 \cdot g_1 + \cdots + n_k \cdot g_k = g$ is usually preferred) this problem is NP-complete (resp., TC^0-complete) if the numbers $g_1, \ldots, g_k, g \in \mathbb{Z}$ are encoded in binary representation [9,11] (resp., unary notation [2]).

In [22], Myasnikov et al. encode elements of the finitely generated group G by words over the group generators and their inverses, which corresponds to the unary encoding of integers. There is also an encoding of words that corresponds to the binary encoding of integers, so called straight-line programs, and knapsack problems under this encoding have been studied in [18]. In this paper, we only consider the case where input words are explicitly represented. Here is a list of known results concerning the knapsack problem:

- Knapsack can be solved in polynomial time for every hyperbolic group [22]. In [4] this result was extended to free products of any finite number of hyperbolic groups and finitely generated abelian groups.

© Springer Nature Switzerland AG 2018
I. Potapov and P.-A. Reynier (Eds.): RP 2018, LNCS 11123, pp. 87–102, 2018.
https://doi.org/10.1007/978-3-030-00250-3_7

– There are nilpotent groups of class 2 for which knapsack is undecidable. Examples are direct products of sufficiently many copies of the discrete Heisenberg group $H_3(\mathbb{Z})$ [12], and free nilpotent groups of class 2 and sufficiently high rank [20].

– Knapsack for $H_3(\mathbb{Z})$ is decidable [12]. In particular, together with the previous point it follows that decidability of knapsack is not preserved under direct products.

– Knapsack is decidable for every co-context-free group [12], i.e., groups where the set of all words over the generators that do not represent the identity is a context-free language. Lehnert and Schweitzer [14] have shown that the Higman-Thompson groups are co-context-free.

– Knapsack belongs to NP for all virtually special groups (finite extensions of subgroups of graph groups) [19]. The class of virtually special groups is very rich. It contains all Coxeter groups, one-relator groups with torsion, fully residually free groups, and fundamental groups of hyperbolic 3-manifolds. For graph groups (also known as right-angled Artin groups) a complete classification of the complexity of knapsack was obtained in [19]: If the underlying graph contains an induced path or cycle on 4 nodes, then knapsack is NP-complete; in all other cases knapsack can be solved in polynomial time (even in LogCFL).

– Decidability of knapsack is preserved by finite extensions, HNN-extensions over finite associated subgroups and amalgamated products over finite subgroups [18].

In this paper we further investigate the knapsack problem in hyperbolic groups. The definition of hyperbolic groups requires that all geodesic triangles in the Cayley-graph are δ-slim for a constant δ; see Sect. 3 for details. The class of hyperbolic groups has several alternative characterizations (e.g., it is the class of finitely generated groups with a linear Dehn function), which gives hyperbolic groups a prominent role in geometric group theory. Moreover, in a certain probabilistic sense, almost all finitely presented groups are hyperbolic [8,23]. Also from a computational viewpoint, hyperbolic groups have nice properties: it is known that the word problem and the conjugacy problem can be solved in linear time [3,10]. As mentioned above, knapsack can be solved in polynomial time for every hyperbolic group [22]. Our first main result of this paper provides a precise characterization of the complexity of knapsack for hyperbolic groups: for every hyperbolic group, knapsack belongs to LogCFL, which is the class of all problems that are logspace-reducible to a context-free language. LogCFL has several alternative characterizations, see Sect. 5 for details. The LogCFL upper bound for knapsack in hyperbolic groups improves the polynomial upper bound shown in [22], and also generalizes a result from [15], stating that the word problem for a hyperbolic group is in LogCFL. For hyperbolic groups that contain a copy of a non-abelian free group (such hyperbolic groups are called non-elementary) it follows from [19] that knapsack is LogCFL-complete. Hyperbolic groups that contain no copy of a non-abelian free group (so called elementary hyperbolic groups) are known to be virtually cyclic, in which case knapsack can be shown to be in $\mathsf{NL} \subseteq \mathsf{LogCFL}$.

In Sect. 6 we prove our second main result: for every hyperbolic group G and every tuple $(g, g_1, \ldots, g_k)$ of elements of G the set of all $(n_1, \ldots, n_k) \in \mathbb{N}^k$ such that $g = g_1^{n_1} \cdots g_k^{n_k}$ in G is effectively semilinear. In other words: the set of all solutions of a knapsack instance in G is semilinear. Groups with this property are also called knapsack-semilinear. For the special case $G = \mathbb{Z}$ this is well-known (the set of solutions of a linear equation is Presburger definable and hence semilinear). Clearly, knapsack is decidable for every knapsack-semilinear group (due to the effectiveness assumption). In a series of recent papers it turned out that the class of knapsack-semilinear groups is surprisingly rich. It contains all virtually special groups [16] and all co-context-free group [12] and is closed under the following constructions: going to a finitely generated subgroup (this is trivial), going to a finite group extension [18], HNN-extensions over finite associated subgroups [18], amalgamated free products over finite subgroups [18], direct products (this follows from the closure of semilinear sets under intersection), and restricted wreath products [5].

Our proof of the knapsack-semilinearity of a hyperbolic group shows an additional quantitative statement: If the group elements $g, g_1, \ldots, g_k$ are represented by words over the generators and the total length of these words is N, then the set $\{(n_1, \ldots, n_k) \in \mathbb{N}^k \mid g = g_1^{n_1} \cdots g_k^{n_k} \text{ in } G\}$ has a semilinear representation, where all vectors only contain integers of size at most $p(N)$. Here, $p(x)$ is a fixed polynomial that only depends on G. Groups with this property are called knapsack-tame in [19]. In [19], it is shown that the class of knapsack-tame groups is closed under free products and direct products with $\mathbb{Z}$.

Missing proofs can be found in the long version [17].

2   General Notations

We assume that the reader is familiar with basic concepts from group theory and formal languages. The empty word is denoted with ε. For a word $w = a_1 a_2 \cdots a_n$ let $|w| = n$ be the length of w, and for $1 \leq i \leq j \leq n$ let $w[i] = a_i$, $w[i : j] = a_i \cdots a_j$, $w[: i] = w[1 : i]$ and $w[i :] = w[i : n]$. Moreover, let $w[i : j] = \varepsilon$ for $i > j$.

We let $\mathbb{N} = \{0, 1, 2, \ldots\}$. A set of vectors $A \subseteq \mathbb{N}^k$ is *linear* if there exist vectors $v_0, \ldots, v_n \in \mathbb{N}^k$ such that $A = \{v_0 + \lambda_1 \cdot v_1 + \cdots + \lambda_n \cdot v_n \mid \lambda_1, \ldots, \lambda_n \in \mathbb{N}\}$. The tuple of vectors $(v_0, \ldots, v_n)$ is a *linear representation* of A. Its *magnitude* is the largest number appearing in one the vectors $v_0, \ldots, v_n$. A set $A \subseteq \mathbb{N}^k$ is *semilinear* if it is a finite union of linear sets $A_1, \ldots, A_m$. A *semilinear representation* of A is a list of linear representations for the linear sets $A_1, \ldots, A_m$. Its *magnitude* is the maximal magnitude of the linear representations for the sets $A_1, \ldots, A_m$. The magnitude of a semilinear set A is the smallest magnitude among all semilinear representations of A.

In the context of knapsack problems, we will consider semilinear subsets as mapping $f \colon \{x_1, \ldots, x_k\} \to \mathbb{N}$ for a finite set of variables $X = \{x_1, \ldots, x_k\}$. Such a mapping f can be identified with the vector $(f(x_1), \ldots, f(x_k))$. This allows to use all vector operations (e.g. addition and scalar multiplication) on the set $\mathbb{N}^X$ of all mappings from X to $\mathbb{N}$. The pointwise product $f \cdot g$ of two mappings $f, g \in \mathbb{N}^X$ is defined by $(f \cdot g)(x) = f(x) \cdot g(x)$ for all $x \in X$. Moreover, for mappings $f \in \mathbb{N}^X$, $g \in \mathbb{N}^Y$ with $X \cap Y = \emptyset$ we define $f \oplus g \colon X \cup Y \to \mathbb{N}$ by $(f \oplus g)(x) = f(x)$ for $x \in X$ and $(f \oplus g)(y) = g(y)$ for $y \in Y$. All operations on $\mathbb{N}^X$ will be extended to subsets of $\mathbb{N}^X$ in the standard pointwise way.

It is well-known that the semilinear subsets of $\mathbb{N}^k$ are exactly the sets definable in *Presburger arithmetic*. These are those sets that can be defined with a first-order formula $\varphi(x_1, \ldots, x_k)$ over the structure $(\mathbb{N}, 0, +, \leq)$ [7]. Moreover, the transformations between such a first-order formula and an equivalent semilinear representation are effective. In particular, the semilinear sets are effectively closed under Boolean operations.

3   Hyperbolic Groups

Let G be a finitely generated group with the finite symmetric generating set Σ, i.e., $a \in \Sigma$ implies that $a^{-1} \in \Sigma$. The *Cayley-graph* of G (with respect to Σ) is the undirected graph $\Gamma = \Gamma(G)$ with node set G and all edges (g, ga) for $g \in G$ and $a \in \Sigma$. We view Γ as a geodesic metric space, where every edge (g, ga) is identified with a unit-length interval. It is convenient to label the directed edge from g to ga with the generator a. The distance between two points p, q is denoted with $d_\Gamma(p, q)$. For $g \in G$ let $|g| = d_\Gamma(1, g)$. For $r \geq 0$, let $\mathcal{B}_r(1) = \{g \in G \mid d_\Gamma(1, g) \leq r\}$.

Paths can be defined in a very general way for metric spaces, but we only need paths that are induced by words over Σ. Given a word $w \in \Sigma^*$ of length n, one obtains a unique path $P[w] \colon [0, n] \to \Gamma$, which is a continuous mapping from the real interval $[0, n]$ to Γ. It maps the subinterval $[i, i+1] \subseteq [0, n]$ isometrically onto the edge (g_i, g_{i+1}) of Γ, where g_i (resp., g_{i+1}) is the group element represented by the word $w[:i]$ (resp., $w[:i+1]$). The path $P[w]$ starts in $1 = g_0$ and ends in g_n (the group element represented by w). We also say that $P[w]$ is the unique path that starts in 1 and is labelled with the word w. More generally, for $g \in G$, we denote with $g \cdot P[w]$ the path that starts in g and is labelled with w. When writing $u \cdot P[w]$ for a word $u \in \Sigma^*$, we mean the path $g \cdot P[w]$, where g is the group element represented by u. A path $P \colon [0, n] \to \Gamma$ of the above form is geodesic if $d_\Gamma(P(0), P(n)) = n$; it is a (λ, ϵ)-*quasigeodesic* if $|a - b| \leq \lambda \cdot d_\Gamma(P(a), P(b)) + \varepsilon$ for all $a, b \in [0, n]$; and it is ζ-*local* (λ, ϵ)-*quasigeodesic* if $|a - b| \leq \lambda \cdot d_\Gamma(P(a), P(b)) + \varepsilon$ for all $a, b \in [0, n]$ with $|a - b| \leq \zeta$.

A word $w \in \Sigma^*$ is geodesic if the path $P[w]$ is geodesic, which means that there is no shorter word representing the same group element from G. Similarly, we define the notion of (ζ-local) (λ, ϵ)-quasigeodesic words. A word $w \in \Sigma^*$ is *shortlex reduced* if it is the length-lexicographically smallest word that represents the same group element as w. For this, we have to fix an arbitrary linear order on Σ. Note that if $u = xy$ is shortlex reduced then x and y are shortlex reduced too. For a word $u \in \Sigma^*$ we denote with $\mathsf{shlex}(u)$ the unique shortlex reduced word with $\mathsf{shlex}(u) = u$ in G.

P_2

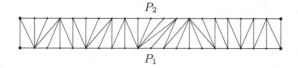

P_1

Fig. 1. Paths that asynchronously K-fellow travel

A *geodesic triangle* consists of three points $p, q, r \in G$ and geodesic paths $P_1 = P_{p,q}$, $P_2 = P_{p,r}$, $P_3 = P_{q,r}$ (the three sides of the triangle), where $P_{x,y}$ is a geodesic path from x to y. The geodesic triangle is δ-*slim* for $\delta \geq 0$, if for all $i \in \{1, 2, 3\}$, every point on P_i has distance at most δ from a point on $\bigcup_{j \in \{1,2,3\} \setminus \{i\}} P_j$. The group G is called δ-*hyperbolic*, if every geodesic triangle is δ-slim. Finally, G is hyperbolic, if it is δ-hyperbolic for some $\delta \geq 0$. Finitely generated free groups are for instance 0-hyperbolic. The property of being hyperbolic is independent of the chosen generating set Σ. The word problem for a hyperbolic group can be solved in real time [10].

Let us fix a δ-hyperbolic group G with the finite symmetric generating set Σ for the rest of the section, and let Γ be the corresponding geodesic metric space. We will apply a couple of well-known results for hyperbolic groups.

Lemma 1 ([6, 8.21]). *Let $g \in G$ be of infinite order and let $n \geq 0$. Let u be a geodesic word representing g. Then the word u^n is (λ, ϵ)-quasigeodesic, where $\lambda = N|g|$, $\epsilon = 2N^2|g|^2 + 2N|g|$ and $N = |\mathcal{B}_{2\delta}(1)|$.*

Consider paths $P_1 \colon [0, n_1] \to \Gamma$, $P_2 \colon [0, n_2] \to \Gamma$ and let K be a positive real number. We say that P_1 and P_2 *asynchronously K-fellow travel* if there exist two continuous non-decreasing mappings $\varphi_1 \colon [0, 1] \to [0, n_1]$ and $\varphi_2 \colon [0, 1] \to [0, n_2]$ such that $\varphi_1(0) = \varphi_2(0) = 0$, $\varphi_1(1) = n_1$, $\varphi_2(1) = n_2$ and for all $0 \leq t \leq 1$, $d_\Gamma(P_1(\varphi_1(t)), P_2(\varphi_2(t))) \leq K$. Intuitively, this means that one can travel along the paths P_1 and P_2 asynchronously with variable speeds such that at any time instant the current points have distance at most K. By slightly increasing K one obtains a ladder graph of the form shown in Fig. 1, where the edges connecting the horizontal P_1- and P_2-labelled paths represent paths of length at most K that connect elements from G.

Lemma 2 ([21]). *Let P_1 and P_2 be (λ, ϵ)-quasigeodesic paths such that P_i starts in g_i and ends in h_i. Assume that $d_\Gamma(g_1, h_1), d_\Gamma(g_2, h_2) \leq h$. There exists a computable bound $K = K(\delta, \lambda, \epsilon, h) \geq h$ such that P_1 and P_2 asynchronously K-fellow travel.*

Finally we need the following lemma, see [17].

Lemma 3. *Fix constants λ, ϵ and let $\kappa = K(\delta, \lambda, \epsilon, 0)$ be taken from Lemma 2. Let $v_1, v_2 \in \Sigma^*$ be geodesic words and $u_1, u_2 \in \Sigma^*$ (λ, ϵ)-quasigeodesic words such that $v_1 u_1 = u_2 v_2$ in G. Consider a factorization $u_1 = x_1 y_1$ with $|x_1| \geq \lambda(|v_1| + 2\delta + \kappa) + \epsilon$ and $|y_1| \geq \lambda(|v_2| + 2\delta + \kappa) + \epsilon$ Then there exists a factorization $u_2 = x_2 y_2$ and $c \in \mathcal{B}_{2\delta + 2\kappa}(1)$ such that $v_1 x_1 = x_2 c$ and $c y_1 = y_2 v_2$ in G.*

4   Knapsack Problems

Let G be a finitely generated group with the finite symmetric generating set Σ. Moreover, let X be a set of variables that take values from $\mathbb{N}$. A *knapsack expression* over G is a formal expression of the form $E = u_1^{x_1} v_1 u_2^{x_2} v_2 \cdots u_k^{x_k} v_k$ with $k \geq 1$, $x_1, \ldots, x_k \in X$, $x_i \neq x_j$ for $i \neq j$, and $u_1, v_1, \ldots, u_k, v_k \in \Sigma^*$. Let $X_E = \{x_1, \ldots, x_k\}$ be the set of variables that occur in E. A solution for E is a mapping $\sigma \in \mathbb{N}^{X_E}$ such that the word $u_1^{\sigma(x_1)} v_1 u_2^{\sigma(x_2)} v_2 \cdots u_k^{\sigma(x_k)} v_k$ represents the identity element of G. With $\mathsf{sol}(E)$ we denote the set of all solutions of E. The *length* of E is defined as $|E| = \sum_{i=1}^{k} |u_i| + |v_i|$, whereas k is its *depth*. The *knapsack problem for* G is the following decision problem: Given a knapsack expression E over G, is $\mathsf{sol}(E)$ non-empty?

The group G is called *knapsack-semilinear* if for every knapsack expression E over G, the set $\mathsf{sol}(E)$ is semilinear and a semilinear representation can be effectively computed from E. The discrete Heisenberg group $H_3(\mathbb{Z})$ (which consists of all upper triangular (3×3)-matrices over the integers, where all diagonal entries are 1) is an example of a group which is not knapsack-semilinear, but for which the knapsack problem is decidable, see [12].

The group G is called *polynomially knapsack-bounded* if there is a fixed polynomial $p(n)$ such that for a given a knapsack expression E over G, one has $\mathsf{sol}(E) \neq \emptyset$ if and only if there exists $\nu \in \mathsf{sol}(E)$ with $\nu(x) \leq p(|E|)$ for all variables x in E. Finally, G is called *knapsack-tame* if there is a fixed polynomial $p(n)$ such that for a given a knapsack expression E over G one can compute a semilinear representation for $\mathsf{sol}(E)$ of magnitude at most $p(|E|)$. Thus, every knapsack-tame group is knapsack-semilinear as well as polynomially knapsack-bounded.

5   Complexity of Knapsack in Hyperbolic Groups

In this section we show that for every hyperbolic group the knapsack problem belongs to the complexity class LogCFL. This class consists of all computational problems that are logspace reducible to a context-free language. The class LogCFL is included in the parallel complexity class NC^2 and has several alternative characterizations; see [24, 26] for details. For our purposes, a characterization via AuxPDAs is most suitable. An AuxPDA (for auxiliary pushdown automaton) is a nondeterministic pushdown automaton with a two-way input tape and an additional work tape. Here we only consider AuxPDAs with the following two restrictions:

- The length of the work tape is restricted to $O(\log n)$ for an input of length n (logspace bounded).
- There is a polynomial $p(n)$, such that every computation path of the AuxPDA on an input of length n has length at most $p(n)$ (polynomially time bounded).

Whenever we speak of an AuxPDA in the following, we implicitly assume that the AuxPDA is logspace bounded and polynomially time bounded. The class of languages that are accepted by such AuxPDAs is exactly LogCFL [24]. A *one-way* AuxPDA is an AuxPDA that never moves the input head to the left. Hence, in every step, the input head either does not move, or moves to the right.

In order to show that knapsack for a hyperbolic group belongs to LogCFL, we use the following important result from [22]:

Theorem 4 ([22]). *Every hyperbolic group is polynomially knapsack-bounded.*

This result is also a direct corollary of Theorem 7 from the next section.

In [15] it is shown that the word problem for a hyperbolic group belongs to LogCFL. Here, we extend the proof from [15] to the knapsack problem. First, we consider another problem of independent interest. An acyclic NFA is a nondeterministic finite automaton $\mathcal{A} = (Q, \Sigma, \Delta, q_0, F)$ (Q is a finite set of states, Σ is the input alphabet, $\Delta \subseteq Q \times \Sigma^* \times Q$ is the set of transition triples, $q_0 \in Q$ is the initial state, and $F \subseteq Q$ is the set of final states) such that the relation $\{(p, q) \in Q \times Q \mid \exists w \in \Sigma^* \colon (p, w, q) \in \Delta\}$ is acyclic. Note that we allow transitions labelled with words; this will be convenient in the proof of the next theorem. For a finitely generated group G with the finite generating set Σ (the concrete choice of Σ is not relevant), the *membership problem for acyclic NFAs over G* is the following computational problem: Given an acyclic NFA $\mathcal{A}$ with input alphabet Σ, does $\mathcal{A}$ accept a word $w \in \Sigma^*$ such that $w = 1$ in G?

Theorem 5. *Membership for acyclic NFAs over a hyperbolic group belongs to* LogCFL.

Proof. Let G be a hyperbolic group with the symmetric generating set Σ and let $\mathcal{A}$ be the input NFA. Let $W = \{w \in \Sigma^* \mid w = 1 \text{ in } G\}$ be the word problem for G. In [15] it is shown that W is a growing context-sensitive language, i.e., it can be generated by a grammar where all productions are strictly length-increasing (except for the start production $S \to \varepsilon$). Hence, by the main result of [1], W can be recognized by a one-way AuxPDA $\mathcal{P}$ in logarithmic space and polynomial time.

An AuxPDA for the membership problem for acyclic NFAs over G guesses a path in the NFA $\mathcal{A}$ and thereby simulates the AuxPDA $\mathcal{P}$ on the word spelled by the guessed path. If the final state of the input NFA $\mathcal{A}$ is reached and the AuxPDA $\mathcal{P}$ accepts at the same time, then the overall AuxPDA accepts. It is important that the AuxPDA $\mathcal{P}$ works one-way since the guessed path in $\mathcal{A}$ cannot be stored in logspace. This implies that the AuxPDA cannot re-access input symbols that have already been processed. The AuxPDA is clearly logspace bounded and polynomially time bounded since $\mathcal{A}$ is acyclic.   □

Theorem 6. *For every hyperbolic groups G, knapsack can be solved in* LogCFL. *Moreover, if G contains a copy of F_2 (the free group of rank 2) then knapsack for G is* LogCFL-*complete.*

Proof. Let G be a hyperbolic group. It is straightforward to present a logspace reduction from knapsack for G to the membership problem for acyclic NFAs.

By Theorem 5, this proves the first statement of the theorem. Consider a knapsack expression $E = u_1^{x_1} v_1 u_2^{x_2} v_2 \cdots u_k^{x_k} v_k$ over G. By Theorem 4, there exists a polynomial $p(x)$ such that $\mathsf{sol}(E) \neq \emptyset$ if and only if there exists a solution $(n_1, \ldots, n_k) \in \mathsf{sol}(E)$ such that $n_i \leq p(|E|)$ for all $1 \leq i \leq k$. For our reduction it therefore suffices to construct an acyclic NFA for the finite language $\{u_1^{n_1} v_1 u_2^{n_2} v_2 \cdots u_k^{n_k} v_k \mid 0 \leq n_1, \ldots, n_k \leq p(|E|)\}$, which is easy (see also [19, Sect. 4.2.5]).

The second statement from the theorem follows from [19, Proposition 4.26], where it was shown that knapsack for F_2 is LogCFL-complete.   □

6   Hyperbolic Groups Are Knapsack-Semilinear

In this section, we prove the following strengthening of Theorem 4:

Theorem 7. *Every hyperbolic group is knapsack-tame.*

Let us remark that the total number of vectors in a semilinear representation can be exponential, even for the simplest case $G = \mathbb{Z}$. Take the (additively written) knapsack expression $E = x_1 + x_2 + \cdots + x_n - n$. Then $\mathsf{sol}(E)$ is finite and consists of $\binom{2n-1}{n} \geq 2^n$ vectors.

Let us fix a δ-hyperbolic group G for the rest of Sect. 6 and let Σ be a finite symmetric generating set for G.

6.1   Knapsack Expressions of Depth Two

We first consider knapsack expressions of depth 2 where all powers are quasi-geodesic. It is well known that the semilinear sets are exactly the Parikh images of the regular languages. We need the following quantitative version of this result:

Theorem 8 ([25, Theorem 4.1], see also [13]). *Let k be a fixed constant. Given an NFA $\mathcal{A}$ over an alphabet of size k with n states, one can compute in polynomial time a semilinear representation of the Parikh image of $L(\mathcal{A})$. Moreover, all numbers appearing in the semilinear representation are polynomially bounded in n.*

Lemma 9. *Let λ and ϵ be fixed constants. For all geodesic words $u_1, v_1, u_2, v_2 \in \Sigma^*$ such that $u_1 \neq \varepsilon \neq u_2$ and u_1^n, u_2^n are (λ, ϵ)-quasigeodesic for all $n \geq 0$, the set $\{(x_1, x_2) \in \mathbb{N} \times \mathbb{N} \mid v_1 u_1^{x_1} = u_2^{x_2} v_2 \text{ in } G\}$ is effectively semilinear with magnitude bounded by $p(|u_1| + |v_1| + |u_2| + |v_2|)$ for a fixed polynomial $p(n)$.*

Proof. Let $S := \{(x_1, x_2) \in \mathbb{N} \times \mathbb{N} \mid v_1 u_1^{x_1} = u_2^{x_2} v_2 \text{ in } G\}$. We will define an NFA $\mathcal{A}$ over the alphabet $\{a_1, a_2\}$ such that the Parikh image of $L(\mathcal{A})$ is S. Moreover, the number of states of $\mathcal{A}$ is polynomial in $|u_1| + |u_2| + |v_1| + |v_2|$. This allows us to apply Theorem 8. We will allow transitions that are labelled with words (having length polynomial in $|u_1| + |u_2| + |v_1| + |v_2|$). Moreover, instead of writing

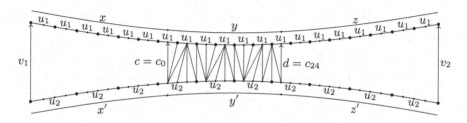

Fig. 2. Example for the construction from the proof of Lemma 9.

in the transitions these words, we write their Parikh images (so, for instance, a transition $p \xrightarrow{a_1 a_2 a_1} q$ is written as $p \xrightarrow{(2,1)} q$.

Let $\ell_i = |u_i|$ and $m_i = |v_i|$. Take the constant κ from Lemma 3 and define $N_1 = \lambda(m_1 + 2\delta + \kappa) + \epsilon$ and $N_2 = \lambda(m_2 + 2\delta + \kappa) + \epsilon$. We split the solution set S into $S_1 = S \cap \{(n_1, n_2) \in \mathbb{N} \times \mathbb{N} \mid n_1 < (N_1 + N_2)/\ell_1\}$ and $S_2 = S \setminus S_1$. For all $(n_1, n_2) \in S_1$ we have $|u_1^{n_1}| = n_1 \ell_1 < N_1 + N_2$. Hence, $|\mathsf{shlex}(u_2^{n_2})| = |\mathsf{shlex}(v_1 u_1^{n_1} v_2^{-1})| < N_1 + N_2 + m_1 + m_2$. Since $u_2^{n_2}$ is (λ, ϵ)-quasigeodesic we get $|u_2^{n_2}| = n_2 \ell_2 < \lambda(N_1 + N_2 + m_1 + m_2) + \epsilon$, i.e., $n_2 < (\lambda(N_1 + N_2 + m_1 + m_2) + \epsilon)/\ell_2$. Hence, S_1 is finite and its magnitude is bounded by $\mathcal{O}(m_1 + m_2)$.

We now deal with pairs $(n_1, n_2) \in S_2$. Consider such a pair (n_1, n_2) and the quasigeodesic rectangle consisting of the four paths $Q_1 = P[v_1]$, $P_1 = v_1 \cdot P[u_1^{n_1}]$, $P_2 = P[u_2^{n_2}]$, and $Q_2 = u_2^{n_2} \cdot P[v_2]$. Since $|u_1^{n_1}| \geq N_1 + N_2$, we factorize the word $u_1^{n_1}$ as $u_1^{n_1} = xyz$ with $|x| = N_1$ and $|z| = N_2$. By Lemma 3 we can factorize $u_2^{n_2}$ as $u_2^{n_2} = x'y'z'$ such that there exist $c, d \in \mathcal{B}_{2\delta+2\kappa}(1)$ with $v_1 x = x'c$ and $dz = z'v_2$ in G, see Fig. 2 (where $n_1 = 20$, $n_2 = 10$, $\ell_1 = 2$ and $\ell_2 = 4$). Since $u_2^{n_2}$ is (λ, ϵ)-quasigeodesic, we have

$$|x'| \leq \lambda(m_1 + |x| + 2\delta + 2\kappa) + \epsilon = \lambda(m_1 + N_1 + 2\delta + 2\kappa) + \epsilon, \tag{1}$$

$$|z'| \leq \lambda(m_2 + |z| + 2\delta + 2\kappa) + \epsilon = \lambda(m_2 + N_2 + 2\delta + 2\kappa) + \epsilon. \tag{2}$$

Consider now the subpath P_1' of P_1 from $P_1(|x|)$ to $P_1(n_1 \ell_1 - |z|)$ and the subpath P_2' of P_2 from $P_2(|x'|)$ to $P_2(n_2 \ell_2 - |z'|)$. These are the paths labelled with y and y', respectively, in Fig. 2. By Lemma 2 these paths asynchronously γ-fellow travel, where $\gamma := K(\delta, \lambda, \epsilon, 2\delta + 2\kappa)$ is a constant. In Fig. 2 this is visualized by the part between the c-labelled edge and the d-labelled edge. W.l.o.g. we assume that $\gamma \geq 2\delta + 2\kappa$.

We now define the NFA $\mathcal{A}$ over the alphabet $\{a_1, a_2\}$ (recall the we replace edge labels from $\{a_1, a_2\}^*$ by their Parikh images). The state set of $\mathcal{A}$ is

$$Q = \{q_0, q_f\} \cup \{(i, b, j) \mid 0 \leq i < \ell_1, 0 \leq j < \ell_2, b \in \mathcal{B}_\gamma(1)\}.$$

The unique initial (resp., final) state is q_0 (resp., q_f). To define the transitions of $\mathcal{A}$ set $p = \lfloor N_1/\ell_1 \rfloor = \lfloor |x|/|u_1| \rfloor$, $r = N_1 \bmod \ell_1 = |x| \bmod |u_1|$, $s = \lceil N_2/\ell_1 \rceil = \lceil |z|/|u_1| \rceil$, $t = -N_2 \bmod \ell_1 = -|z| \bmod |u_1|$. Thus, we have $x = u_1^p u_1[: r]$ and $z = u_1^s[t + 1 :]$. There are the following types of transitions (transitions without a label are implicitly labelled by the zero vector $(0,0)$), where $0 \leq i < \ell_1$, $0 \leq j < \ell_2$, $b, b' \in \mathcal{B}_\gamma(1)$.

1. $q_0 \xrightarrow{(p,p')} (r, c, r')$ if there exists a number $0 \leq k \leq \lambda(m_1 + N_1 + 2\delta + 2\kappa) + \epsilon$ (this is the possible range for the length of x' in (1)) such that $p' = \lfloor k/\ell_2 \rfloor$, $r' = k \bmod \ell_2$, and $v_1 u_1^p u_1[: r] = u_2^{p'} u_2[: r']c$ in G.
2. $(i, b, j) \rightarrow (i+1, b', j)$ if $i+1 < \ell_1$ and $bu_1[i+1] = b'$ in G.
3. $(\ell_1 - 1, b, j) \xrightarrow{(1,0)} (0, b', j)$ if $bu_1[\ell_1] = b'$ in G.
4. $(i, b, j) \rightarrow (i, b', j+1)$ if $j+1 < \ell_2$ and $b = u_2[j+1]b'$ in G.
5. $(i, b, \ell_2 - 1) \xrightarrow{(0,1)} (i, b', 0)$ if $b = u_2[\ell_2]b'$ in G.
6. $(t, d, t') \xrightarrow{(s,s')} q_f$ if there exists a number $0 \leq k \leq \lambda(m_2 + N_2 + 2\delta + 2\kappa) + \epsilon$ (this is the possible range for the length of z' in (2)) such that $s' = \lceil k/\ell_2 \rceil$, $t' = -k \bmod \ell_2$, and $du_1[t+1 :]u_1^s = u_2[t'+1 :]u_1^s v_2$ in G.

The construction is best explained using the example in Fig. 2. As mentioned above, the vertical lines between $c = c_0$ and $d = c_{24}$ represent the asynchronous γ-fellow travelling. The vertical lines are labelled with group elements $c_0, c_1, \ldots, c_{23}, c_{24} \in \mathcal{B}_\gamma(1)$ from left to right. In order to not overload the figure we only show c_0 and c_{24}. Note that $x = u_1^6 u_1[1]$, $x' = u_2^3 u_2[1]$, $z = u_1[2]u_1^7$, $z' = u_2[2 : 4]u_2^3$. Basically, the NFA $\mathcal{A}$ moves the vertical edges from left to right and thereby stores (i) the label c_i of the vertical edge, (ii) the position in the current u_2-factor where the vertical edge starts (position 0 means that we have just completed a u_2-factor), and (iii) the position in the current u_1-factor where the vertical edge ends. If a u_1-factor (resp., u_2-factor) is completed then the automaton makes a $(1,0)$-labelled (resp., $(0,1)$-labelled) transition. The complete run that corresponds to Fig. 2 is:

$$q_0 \xrightarrow{(6,3)} (1, c_0, 1) \xrightarrow{(1,0)} (0, c_1, 1) \rightarrow (1, c_2, 1) \rightarrow (2, c_3, 1) \rightarrow$$
$$(3, c_4, 1) \xrightarrow{(0,1)} (0, c_5, 1) \xrightarrow{(1,0)} (0, c_6, 0) \rightarrow (0, c_7, 1) \rightarrow$$
$$(1, c_8, 1) \xrightarrow{(1,0)} (0, c_9, 1) \rightarrow (1, c_{10}, 1) \rightarrow (1, c_{11}, 2) \rightarrow$$
$$(1, c_{12}, 3) \xrightarrow{(0,1)} (1, c_{13}, 0) \xrightarrow{(1,0)} (0, c_{14}, 0) \rightarrow (0, c_{15}, 1) \rightarrow$$
$$(1, c_{16}, 1) \xrightarrow{(1,0)} (0, c_{17}, 1) \rightarrow (1, c_{18}, 1) \rightarrow (1, c_{19}, 2) \rightarrow$$
$$(1, c_{20}, 3) \xrightarrow{(1,0)} (0, c_{21}, 3) \xrightarrow{(0,1)} (0, c_{22}, 0) \rightarrow (0, c_{23}, 1) \rightarrow$$
$$(1, c_{24}, 1) \xrightarrow{(8,4)} q_f$$

With the above intuition it is straightforward to show that the Parikh image of $L(\mathcal{A})$ is indeed S_2. Also note that the number of states of $\mathcal{A}$ is bounded by $\mathcal{O}(\ell_1 \ell_2)$. The statement of the lemma then follows directly from Theorem 8. $\square$

6.2 Reduction to Quasi-geodesic Knapsack Expressions

Let us call a knapsack expression $E = u_1^{x_1} v_1 u_2^{x_2} v_2 \cdots u_k^{x_k} v_k$ over G (λ, ϵ)-quasi-geodesic if all $u_1, \ldots, u_k, v_1, \ldots, v_k$ are geodesic and for all $1 \leq i \leq k$ and all

$n \geq 0$ the word u_i^n is (λ, ϵ)-quasigeodesic. We say that E has *infinite order*, if all u_i represent group elements of infinite order. The goal of this section is to reduce a knapsack expression to a finite number (in fact, exponentially many) of (λ, ϵ)-quasigeodesic knapsack expressions of infinite order for certain constants λ, ϵ:

Proposition 10. *There are fixed constants λ, ϵ such that from a given knapsack expression E over G one can compute a finite list of knapsack expressions E_i $(i \in I)$ over G such that*

- $\mathsf{sol}(E) = \bigcup_{i \in I} \left((m_i \cdot \mathsf{sol}(E_i) + d_i) \oplus \mathcal{F}_i \right)$,
- *every $\mathcal{F}_i$ is a semilinear subset of $\mathbb{N}^Y$ for a subset $Y \subseteq X_E$,*
- *the magnitude of every $\mathcal{F}_i$ is bounded by a constant that only depends on G,*
- *every E_i is a (λ, ϵ)-quasigeodesic knapsack expression of infinite order with variables from $Z := X_E \setminus Y$,*
- *the size of every E_i is bounded by $\mathcal{O}(|E|)$, and*
- *all m_i and d_i are vectors from $\mathbb{N}^Z$ where all entries are bounded by a constant that only depends on G (here, $m_i \cdot \mathsf{sol}(E_i) = \{m_i \cdot z \mid z \in \mathsf{sol}(E)\}$ and $m_i \cdot z$ is the pointwise multiplication of the vectors m_i and z).*

Once Proposition 10 is shown, we can conclude the proof of Theorem 7 by showing that all sets $\mathsf{sol}(E_i)$ are semilinear and that their magnitudes are bounded by $p(|E_i|)$ for a fixed polynomial $p(n)$. This will be achieved in the next section.

A detailed proof of Proposition 10 can be found in the long version [17]; here we only provide a sketch. Consider a knapsack expression $E = u_1^{x_1} v_1 u_2^{x_2} v_2 \cdots u_k^{x_k} v_k$. We can assume that every u_i is shortlex reduced. Let $g_i \in G$ be the group element represented by the word u_i. Reducing to the case, where all g_i have infinite order is relatively easy. In a hyperbolic group G the order of torsion elements is bounded by a fixed constant that only depends on G, see also the proof of [22, Theorem 6.7]). This allows to check for each g_i whether it has finite order, and to compute the order in the positive case. Let $Y \subseteq \{x_1, \ldots, x_k\}$ be those variables x_i such that g_i has finite order. For $x_i \in Y$ let $o_i < \infty$ be the order of g_i. Let $\mathcal{F}$ be the set of mappings $f \colon Y \to \mathbb{N}$ such that $0 \leq f(x_i) < o_i$ for all $x_i \in Y$. For every such mapping $f \in \mathcal{F}$ let E_f be the knapsack expression that is obtained from E by replacing for every $x_i \in Y$ the power $u_i^{x_i}$ by $u_i^{f(x_i)}$ (which is merged with the word v_i). Moreover, let $\mathcal{F}_f$ be the set of all mappings $g \colon Y \to \mathbb{N}$ such that $g(x_i) \equiv f(x_i) \bmod o_i$ for every $x_i \in Y$. Then the set $\mathsf{sol}(E)$ can be written as $\mathsf{sol}(E) = \bigcup_{f \in \mathcal{F}} \mathsf{sol}(E_f) \oplus \mathcal{F}_f$. Note that $\mathcal{F}_f$ is a semilinear set of magnitude $\mathcal{O}(1)$.

In a second step we reduce every E_f (which has infinite order) to (λ, ϵ)-quasigeodesic knapsack expressions for fixed constants λ and ϵ. Let us again write $E_f = u_1^{x_1} v_1 u_2^{x_2} v_2 \cdots u_k^{x_k} v_k$. We first use Lemma 1, which tells us that for every $n \geq 0$ and $1 \leq i \leq k$, the word u_i^n is (λ_i, ϵ_i)-quasigeodesic for $\lambda_i = N|u_i|$, $\epsilon_i = 2N^2|u_i|^2 + 2N|u_i|$. In order to reduce these λ_i, ϵ_i to fixed constants we mainly use the following two results from [3], where $L = 34\delta + 2$ and $K = |\mathcal{B}_{4\delta}(1)|^2$ (these are constants):

- Let $u = u_1 u_2$ be shortlex reduced, where $|u_1| \leq |u_2| \leq |u_1| + 1$, and $\tilde{u} =$ shlex$(u_2 u_1)$. If $|\tilde{u}| \geq 2L + 1$ then for every $n \geq 0$, the word $\tilde{u}^n$ is L-local $(1, 2\delta)$-quasigeodesic [3, Lemma 3.1].
- Let u be geodesic such that $|u| \geq 2L + 1$ and for every $n \geq 0$, the word u^n is L-local $(1, 2\delta)$-quasigeodesic. Then one can compute $c \in \mathcal{B}_{4\delta}(1)$ and $1 \leq m \leq K$ such that $(\text{shlex}(c^{-1} u^m c))^n$ is geodesic for all $n \geq 0$ [3, Sect. 3.2].   □

6.3   Proof of Theorem 7

In this subsection we sketch the proof of Theorem 7; a detailed proof can be found in the full version [17]. Consider a knapsack expression $E = u_1^{x_1} v_1 u_2^{x_2} v_2 \cdots u_k^{x_k} v_k$. We can assume that all u_i, v_i are geodesic. By Proposition 10 we can moreover assume that for all $1 \leq i \leq k$, u_i represents a group element of infinite order and that u_i^n is (λ, ϵ)-quasigeodesics for all $n \geq 0$, where λ, ϵ are fixed constants. We want to show that sol(E) is semilinear and has a magnitude that is polynomially bounded by $|E|$.

For the case $k = 1$ we have to consider all $n \in \mathbb{N}$ with $u_1^n = v_1^{-1}$ in G. Since u_1 represents a group element of finite order there is at most one such n. Moreover, since u_i^n is (λ, ϵ)-quasigeodesic, such an n has to satisfy $|u_1| \cdot n \leq \lambda |v_1| + \epsilon$, which yields a linear bound on n. For the case $k = 2$ we can directly use Proposition 9. Now assume that $k \geq 3$. We want to show that the set sol(E) is a semilinear subset of $\mathbb{N}^k$ (later we will consider the magnitude of sol(E)). For this we construct a Presburger formula with free variables $x_1, \ldots, x_k$ that is equivalent to $E = 1$. We do this by induction on the depth k. Therefore, we can use in our Presburger formula also knapsack equations of the form $F = 1$, where F has depth at most $k - 1$. One can also easily observe that it suffices to construct a Presburger formula for sol$(E) \cap (\mathbb{N} \setminus \{0\})^k$.

Consider a tuple $(n_1, \ldots, n_k) \in \text{sol}(E) \cap (\mathbb{N} \setminus \{0\})^k$ and the corresponding $2k$-gon that is defined by the (λ, ϵ)-quasigeodesic paths $P_i = (u_1^{n_1} v_1 \cdots u_{i-1}^{n_{i-1}} v_{i-1}) \cdot P[u_i^{n_i}]$ and the geodesic paths $Q_i = (u_1^{n_1} v_1 \cdots u_i^{n_i}) \cdot P[v_i]$, see Fig. 3a for the case $k = 3$. Since all paths P_i and Q_i are (λ, ϵ)-quasigeodesic, we can apply [22, Lemma 6.4]: Every side of the $2k$-gon is contained in the h-neighborhoods of the other sides, where $h = \xi + \xi \log(2k)$ for a constant ξ that only depends on the constants $\delta, \lambda, \varepsilon$.

Let us now consider the side P_2 of the quasigeodesic $(2k)$-gon. It is labelled with $u_2^{x_2}$. Every point on P_2 must have distance at most h from one of the sides $P_1, Q_1, Q_2, P_3, \ldots, P_k, Q_k$. We distinguish several cases. In each case we cut the $2k$-gon into smaller pieces along paths of length $\leq 2h + 1$ (in fact, length h except for one case), and these smaller pieces will correspond to knapsack expressions of depth $< k$. This is done until all knapsack expressions have depth at most two. Let us consider one typical case, the other cases are considered in the long version [17].

Assume that there is a point $p \in P_2$ that has distance at most h from a point $q \in Q_i$, where $3 \leq i \leq k$. The situation looks as shown in Fig. 3b. For every tuple $t = (w, u_{2,1}, u_{2,2}, v_{i,1}, v_{i,2})$ such that $w \in \Sigma^*$ is of length at most h, $u_2 = u_{2,1} u_{2,2}$ and $v_i = v_{i,1} v_{i,2}$, we construct two new knapsack expressions $F_t =$

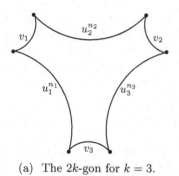

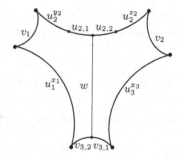

(a) The $2k$-gon for $k = 3$.   (b) Splitting the $2k$-gon into two parts.

Fig. 3. Planar diagrams from the proof of Theorem 7.

$u_1^{x_1} v_1 u_2^{y_2} (u_{2,1} w v_{i,2}) u_{i+1}^{x_{i+1}} v_{i+1} \cdots u_k^{x_k} v_k$, $G_t = u_{2,2} u_2^{z_2} v_2 u_3^{x_3} v_3 \cdots u_i^{x_i} (v_{i,1} w^{-1})$ and the formula

$$\bigvee_t \exists y_2, z_2 : x_2 = y_2 + 1 + z_2 \wedge F_t = 1 \wedge G_t = 1, \tag{3}$$

where t ranges over all tuples of the above form. Here y_2, z_2, y_i, z_i are new variables. Note that F_t and G_t have depth at most $k - 1$.

There are several other cases in which we can similarly split E into several (at most three) knapsack expressions of depth $<k$. In each case, we get a formula similar to (3), and we take the disjunction of all these formulas. This shows that $\mathsf{sol}(E)$ is semilinear.

It remains to argue that the magnitude of $\mathsf{sol}(E)$ is bounded polynomially in $|E|$. Iterating the splitting procedure results in a disjunction of formulas of the form

$$\exists y_1, \ldots, y_m \bigwedge_{i \in I} E_i = 1 \bigwedge_{j \in J} z_j = z_j' + z_j'' + 1, \tag{4}$$

where every E_i is a knapsack expression of depth at most two. Moreover, for $i \neq j$, E_i and E_j have no common variables. The existentially quantified variables $y_1, \ldots, y_m$ are the new variables that were introduced when splitting factors $u_i^{x_i}$ (e.g., y_2, z_2 in (3)). The variables z_j, z_j', z_j'' in (4) are from $\{x_1, \ldots, x_k, y_1, \ldots, y_m\}$. The equations $z_j = z_j' + z_j'' + 1$ in (4) result from the splitting of factors $u_i^{x_i}$. For instance, $x_2 = y_2 + 1 + z_2$ in (3) is one such equation.

In order to bound the magnitude of $\mathsf{sol}(E)$ it suffices to consider a single conjunctive formula of the form (4), since disjunction corresponds to union of semilinear sets, which does not increase the magnitude. We can also ignore the existential quantifiers in (4), because existential quantification corresponds to projection onto some of the coordinates, which cannot increase the magnitude. Hence, we have to consider the magnitude of the semilinear set A defined by the subformula $\bigwedge_{i \in I} E_i = 1 \bigwedge_{j \in J} z_j = z_j' + z_j'' + 1$ of (4). To bound the magnitude of A, we show that (i) the size of every E_i in (4) is bounded by $\mathcal{O}(|E|^2)$ and (ii) that the size of the index set I is bounded by $\mathcal{O}(k^2)$. From (i) it follows that the magnitude of every set $\mathsf{sol}(E_i)$ is bounded polynomially in $|E|$. For the additional variables that are defined by the equations $z_j = z'_j + z''_j + 1$ in (4) one has to notice that these equations $z_j = z'_j + z''_j + 1$ result in a tree-shaped additive circuit whose input gates are the variables that appear in the E_i $(i \in I)$. By (ii) this circuit has $\mathcal{O}(k^2)$ input gates. From this, one can finally deduce that the magnitude of the set A is indeed polynomially bounded in E.   □

7   More Groups with Knapsack in LogCFL

Let $\mathcal{C}$ be the smallest class of groups such that (i) every hyperbolic group belongs to $\mathcal{C}$, (ii) if $G \in \mathcal{C}$ then also $G \times \mathbb{Z} \in \mathcal{C}$, and (iii) if $G, H \in \mathcal{C}$ then also $G * H \in \mathcal{C}$ (where $G * H$ is the free product of G and H). From Theorem 7 and [19, Propositions 4.11 and 4.17] it follows that every group $G \in \mathcal{C}$ is knapsack-tame and hence polynomially knapsack-bounded. Hence, knapsack for a group $G \in \mathcal{C}$ is logspace reducible to membership for acyclic NFAs over G (the reduction in the proof of Theorem 6 works for any group). Finally, it was shown in the full version [17] that the word problem for every group in $\mathcal{C}$ can be accepted by a one-way AuxPDA in logarithmic space and polynomial time (the proof is essentially the same as in [19, Lemma 4.8]). This allows to generalize the proof of Theorem 5 to groups from $\mathcal{C}$. Hence, for every group $G \in \mathcal{C}$, membership for acyclic NFAs over G and knapsack for G can be solved in LogCFL.

8   Conclusion

In this paper, it is shown that every hyperbolic group is knapsack-tame and that the knapsack problem can be solved in LogCFL. Here is a list of open problems that one might consider for future work.

- For the following important groups, it is not known whether the knapsack problem is decidable: braid groups B_n (with $n \geq 3$), solvable Baumslag-Solitar groups $BS_{1,p} = \langle a, t \mid t^{-1}at = a^p \rangle$ (with $p \geq 2$), and automatic groups which are not in any of the known classes with a decidable knapsack problem.
- In [12], it was shown that knapsack is decidable for every co-context-free group. The algorithm from [12] has an exponential running time. Is there a more efficient solution?
- Is there a polynomially knapsack-bounded group which is not knapsack-tame?

Acknowledgement. This work has been supported by the DFG research project LO 748/13-1.

References

1. Buntrock, G., Otto, F.: Growing context-sensitive languages and Church-Rosser languages. Inf. Comput. **141**, 1–36 (1998)
2. Elberfeld, M., Jakoby, A., Tantau, T.: Algorithmic meta theorems for circuit classes of constant and logarithmic depth. Electron. Colloq. Comput. Complex. (ECCC) **18**, 128 (2011)

3. Epstein, D.B.A., Holt, D.F.: The linearity of the conjugacy problem in word-hyperbolic groups. Int. J. Algebra Comput. **16**(2), 287–306 (2006)

4. Frenkel, E., Nikolaev, A., Ushakov, A.: Knapsack problems in products of groups. J. Symb. Comput. **74**, 96–108 (2016)

5. Ganardi, M., König, D., Lohrey, M., Zetzsche, G.: Knapsack problems for wreath products. In: Proceedings of STACS 2018. LIPIcs, vol. 96, pp. 32:1–32:13. Schloss Dagstuhl - Leibniz-Zentrum für Informatik (2018)

6. Ghys, E., de La Harpe, P.: Sur les groupes hyperboliques d'après Mikhael Gromov. Progress in mathematics. Birkhäuser, Basel (1990)

7. Ginsburg, S., Spanier, E.H.: Semigroups, Presburger formulas, and languages. Pac. J. Math. **16**(2), 285–296 (1966)

8. Gromov, M.: Hyperbolic groups. In: Gersten, S.M. (ed.) Essays in Group Theory. MSRI, vol. 8, pp. 75–263. Springer, New York (1987). https://doi.org/10.1007/978-1-4613-9586-7_3

9. Haase, C.: On the complexity of model checking counter automata. Ph.D. thesis, University of Oxford, St Catherine's College (2011)

10. Holt, D.F.: Word-hyperbolic groups have real-time word problem. Int. J. Algebra Comput. **10**, 221–228 (2000)

11. Karp, R.M.: Reducibility among combinatorial problems. In: Miller, R.E., Thatcher, J.W. (eds.) Complexity of Computer Computations, pp. 85–103. Plenum Press, New York (1972)

12. König, D., Lohrey, M., Zetzsche, G.: Knapsack and subset sum problems in nilpotent, polycyclic, and co-context-free groups. In: Algebra and Computer Science. Contemporary Mathematics, vol. 677, pp. 138–153. American Mathematical Society (2016)

13. Kopczynski, E., To, A.W.: Parikh images of grammars: complexity and applications. In: Proceedings of LICS 2010, pp. 80–89. IEEE Computer Society (2010)

14. Lehnert, J., Schweitzer, P.: The co-word problem for the Higman-Thompson group is context-free. Bull. Lond. Math. Soc. **39**(2), 235–241 (2007)

15. Lohrey, M.: Decidability and complexity in automatic monoids. Int. J. Found. Comput. Sci. **16**(4), 707–722 (2005)

16. Lohrey, M., Zetzsche, G.: Knapsack in graph groups, HNN-extensions and amalgamated products. CoRR, abs/1509.05957 (2015)

17. Lohrey, M.: Knapsack in hyperbolic groups. CoRR, abs/1807.06774 (2018). https://arxiv.org/abs/1807.06774

18. Lohrey, M., Zetzsche, G.: Knapsack in graph groups, HNN-extensions and amalgamated products. In: Proceedings of STACS 2016. LIPIcs, vol. 47, pp. 50:1–50:14. Schloss Dagstuhl - Leibniz-Zentrum für Informatik (2016)

19. Lohrey, M., Zetzsche, G.: Knapsack in graph groups. Theory Comput. Syst. **62**(1), 192–246 (2018)

20. Mishchenko, A., Treier, A.: Knapsack problem for nilpotent groups. Groups Complex. Cryptol. **9**(1), 87–98 (2017)

21. Myasnikov, A., Nikolaev, A.: Verbal subgroups of hyperbolic groups have infinite width. J. Lond. Math. Soc. **90**(2), 573–591 (2014)

22. Myasnikov, A., Nikolaev, A., Ushakov, A.: Knapsack problems in groups. Math. Comput. **84**, 987–1016 (2015)

23. Ol'shanskii, A.Yu.: Almost every group is hyperbolic. Int. J. Algebra Comput. **2**(1), 1–17 (1992)

24. Sudborough, I.H.: On the tape complexity of deterministic context-free languages. J. ACM **25**(3), 405–414 (1978)

25. To, A.W.: Parikh images of regular languages: complexity and applications. CoRR, abs/1002.1464 (2010). http://arxiv.org/abs/1002.1464
26. Vollmer, H.: Introduction to Circuit Complexity. Springer, Heidelberg (1999). https://doi.org/10.1007/978-3-662-03927-4

Generalized Tag Systems

Turlough Neary$^{(\boxtimes)}$ and Matthew Cook

Institute of Neuroinformatics, University of Zürich and ETH Zürich,
Zürich, Switzerland
tneary@ini.phys.ethz.ch

Abstract. Tag systems and cyclic tag systems are forms of rewriting systems which, due to the simplicity of their rewrite rules, have become popular targets for reductions when proving universality/undecidability results. They have been used to prove such results for the smallest universal Turing machines, the elementary cellular automata Rule 110, for simple instances of the Post correspondence problem and related problems on simple matrix semi-groups, and many other simple systems. In this work we compare the computational power of tag systems, cyclic tag systems and a straightforward generalization of these two types of rewriting system. We explore the relationships between the various systems by showing that some variants simulate each other in linear time via simple encodings, and that linear time simulations between other variants are not possible using such simple encodings. We also give a cyclic tag system that has only four instructions and simulates repeated iteration of the Collatz function.

1 Introduction

Tag systems were created in 1920 (and published in 1943) by Post [17] to explore the intractability of simple term rewriting systems. A β-tag system operates on a string W by reading the leftmost symbol from W and appending a word (that depends on the read symbol) onto the rightmost end of the W while also deleting the leftmost β symbols from W. The same process is then repeated on the resulting string and so on. It has long been known that simple tag systems exhibit complex behaviour. In fact Post's 3-tag system (read 0 append 00, read 1 append 1101) [17], has given rise to a question that remains intractable to this day: Given an arbitrary binary string will Post's tag system enter a periodic cycle or halt by producing a word of length <3?

Cyclic tag systems were created in 1994 (and published in 2004) by Cook [4] as a stepping stone in proving that the cellular automata Rule 110 is Turing universal. The simple type of circular control flow used by cyclic tag systems allowed Cook to simulate the list of appendants (binary words) that define a cyclic tag system's program as a repeated periodic pattern in a Rule 110 configuration. Simple cyclic tag systems are also known to exhibit complex behaviour:

This work is supported by Swiss National Science Foundation grant numbers 200021-153295 and 200021-166231.

I. Potapov and P.-A. Reynier (Eds.): RP 2018, LNCS 11123, pp. 103–116, 2018.
https://doi.org/10.1007/978-3-030-00250-3_8

The cyclic tag system $(111, 0)$ of Cook remains intractable, regarding whether periodicity ever arises in the infinite sequence for which when we consider all the indices of the positions of the ones, and replace the odd indices with 111 and the even indices with 0, and prepend a single 1, then we get the original infinite sequence.

Both of these systems are special cases of a formalism we introduce in this paper, which we call *generalized tag systems*. Like tag systems and cyclic tag systems, generalized tag systems also operate on a string of symbols, reading from the beginning, and appending to the end based on what is read. At every time step t, they remove the first symbol σ from the string, and append $f_{t \bmod \beta}(\sigma)$ to the end of the string. The function $f_i(\sigma)$ is defined for $i \in 0, .., \beta - 1$ and $\sigma \in \Sigma$, where Σ is the alphabet used for the string, and β is a positive integer that defines the 'cycle', the cyclic period of the system.

Tag systems correspond to the special case of generalized tag systems where $f_i(\sigma)$ yields the empty string for all symbols σ whenever $i > 0$. Cyclic tag systems correspond to the special case of generalized tag systems where $\Sigma = \{0, 1\}$, and $f_i(0)$ yields the empty string for all i.

In this paper we will present each of these systems, and we give straightforward reductions between them, showing that they can simulate each other in linear time. The reductions between tag systems and generalized tag systems keep the same 'cycle'. It appears to be very difficult for such systems with cycle lengths that are relative prime to simulate each other in linear time. We give a theorem showing that any such simulation cannot work in linear time if it uses a direct encoding of the string being operated on. Previous results where the time efficiency of tag systems and cyclic tag systems is explored are to be found in [14, 16, 19].

Various authors have given universality results [3, 4, 15, 18] for tag systems and cyclic tag systems and while DeMol [7] and Cook [5] have both given decidability results, with DeMol showing that the reachability problem is decidable for 2-tag systems with 2 symbols, and Cook showing that the reachability problem is decidable for non-deterministic 1-tag systems, there remains a lot of unexplored space between the simplest universal tag systems and known decidability results. Due to the well documented difficulty of the Collatz Problem [8–10] it is often implemented as instances of models that lie between known decidability upper and lower bounds thereby indicating that decidability results would be difficult to find for such restrictions of the model (see for example [1, 2, 11–13]). DeMol [6] has given a 3-symbol tag system with deletion number 2 that simulates iterations of the Collatz function. Here we give a simple generalized tag system with only three non-blank instructions and a cyclic tag system with only four instructions both of which simulate iterations of the Collatz function.

2   Preliminaries

We write $c_1 \vdash c_2$ if a configuration c_2 is obtained from c_1 via a single computation step. We let $c_1 \vdash^t c_2$ denote a sequence of t computation steps. The length of a word W is denoted $|W|$, and ϵ denotes the empty word. We let $\langle v \rangle$ denote the encoding of v, where v is a symbol or a word. We use the binary modulo operation $a = m \bmod n$, where $a = m - ny$, $0 \leqslant a < n$, and a, m, n, and y are integers.

The well known Collatz function [9,10] is given by

$$f(x) = \begin{cases} \frac{x}{2} & \text{if } 0 = x \bmod 2 \\ \frac{3x+1}{2} & \text{if } 1 = x \bmod 2, \end{cases} \tag{1}$$

2.1 Generalized Tag Systems

Here we introduce generalized tag systems which are a straightforward generalization of both tag systems and cyclic tag systems.

Definition 1. *A generalized tag system consists of a finite alphabet of symbols Σ, a possibly empty set of halt symbols H disjoint from Σ, a cycle number $\beta \in \mathbb{N}$ with $\beta \geqslant 1$, and a set of rules $R : k \times \Sigma \to \{\Sigma \cup H\}^*$ where $0 \leq k < \beta$ and there is exactly one rule for each (k, Σ) pair.*

We call the word on the right hand side of each rule in R an appendant. It is convenient to represent the program of a generalized tag system as a table of appendants as shown in Table 1.

Table 1. Appendants for a generalized tag system with alphabet $\Sigma = \{\sigma_1, \sigma_2, \cdots, \sigma_m\}$, cycle length β, and rules of the form $(k, \sigma_i) \to A_{k,i}$ where $A_{k,i} \in \{\Sigma \cup H\}^*$

k	Σ				
	σ_1	σ_2	σ_3	$\cdots$	σ_m
0	$A_{0,1}$	$A_{0,2}$	$A_{0,3}$	$\cdots$	$A_{0,m}$
1	$A_{1,1}$	$A_{1,2}$	$A_{1,3}$	$\cdots$	$A_{1,m}$
2	$A_{2,1}$	$A_{2,2}$	$A_{2,3}$	$\cdots$	$A_{2,m}$
$\vdots$	$\vdots$	$\vdots$	$\vdots$	$\cdots$	$\vdots$
$\beta-1$	$A_{\beta-1,1}$	$A_{\beta-1,2}$	$A_{\beta-1,3}$	$\cdots$	$A_{\beta-1,m}$

A *configuration* of a generalized tag system is a pair (k, W) where k is a number that points to a row in the table of appendants and $W = w_1 w_2 \ldots w_n$ (where each $w_j \in \Sigma \cup H$) is a word which we call the dataword. In a configuration where $w_1 = \sigma_i \in \Sigma$, the rule for $(k, \sigma_i) \to A_{k,i}$ is applied by appending $A_{k,i}$ to the right end of W, σ_i is deleted from W, and k is incremented to give $(k+1) \bmod \beta$. So a *computation step* of a generalized tag system is deterministic and is given by:

$$(k, \sigma_i w_2 \ldots w_n) \quad \vdash \quad ((k+1) \bmod \beta, w_2 \ldots w_n A_{k,i})$$

A generalized tag system halts if $w_1 \in H$ (i.e. the leftmost symbol of the dataword W is a halt symbol), or if W is the empty word. In an initial configuration of a generalized tag system $k = 0$ and $W \in \Sigma^*$ is the input word. The generalized tag system given in Table 2 computes iterations of the Collatz function: given a configuration of the form $(0, (bb)^x)$ it computes $(0, (bb)^{f(x)})$ in two passes over the dataword. Below is an example of this system started on $(0, (bb)^3)$ computing $f(3) = 5$ followed by $f(5) = 8$ and $f(8) = 4$.

$$
\begin{array}{llllllll}
& (0, b^6) & \vdash & (1, b^5 cbcc) & \vdash & (2, b^4 cbcc) & \vdash & (3, b^3 cbcc) \\
\vdash & (0, b^2 cbcc) & \vdash^2 & (2, (cbcc)^2) & \vdash^4 & (2, cbccb^6) & \vdash^4 & (2, b^{12}) \\
\vdash^2 & (0, b^{10}) & \vdash^{24} & (0, b^{16}) & \vdash^{32} & (0, b^8)
\end{array}
$$

For an arbitrary value x the system in Table 2 computes an iteration of the Collatz function as follows:

Table 2. Generalized tag system that simulates iterations of the Collatz function

	b	c
0	$cbcc$	bb
1	ϵ	$bbbb$
2	ϵ	ϵ
3	ϵ	ϵ

$$
\begin{array}{lllll}
(0, (bb)^x) & \vdash^{2x} & (0, (cbcc)^{\frac{x}{2}}) & \vdash^{2x} & (0, (bb)^{\frac{x}{2}}) \qquad\qquad \text{if } 0 = x \bmod 2
\end{array}
$$

$$
(0, (bb)^x) \;\; \vdash^{2x} \;\; (2, (cbcc)^{\frac{x+1}{2}}) \;\; \vdash^{2x+2} \;\; (2, (bb)^{\frac{3x+3}{2}}) \;\; \vdash^2 \;\; (0, (bb)^{\frac{3x+1}{2}})
$$
$$
\text{if } 1 = x \bmod 2
$$

During the first pass for each $(bb)^2$ we cyclic through the program once appending a single $cbcc$ thus producing $(cbcc)^{\frac{x}{2}}$, and if x is odd the final bb pair appends an $cbcc$ without completing the final cycle giving $(cbcc)^{\frac{x+1}{2}}$. If x is even we begin the second pass at program row 0 and so each $cbcc$ that is read appends bb giving $(bb)^{\frac{x}{2}}$. If x is odd we begin the second pass at program row 2 and so each $cbcc$ that is read appends $(bb)^3$ giving $(bb)^{\frac{3x+3}{2}}$ and after the last $cbcc$ is read we take 2 further steps to move program control to row 0 by reading a bb pair giving $(bb)^{\frac{3x+1}{2}}$.

2.2 Restrictions of Generalized Tag Systems

2.2.1 Tag Systems

Definition 2. *A* tag system *consists of a finite set of symbols Σ, a finite set of halt symbols H disjoint from Σ, a set of rules $R : \Sigma \to \{\Sigma \cup H\}^*$ with exactly one rule for each element of Σ, and a deletion number $\beta \in \mathbb{N}$ with $\beta \geqslant 1$.*

A tag system configuration is a word $W = w_1 w_2 \ldots w_n$ (here each $w_i \in \Sigma \cup H$). In a configuration where $w_1 = \sigma_i \in \Sigma$ the symbols $\sigma_i w_2 w_3 \ldots w_\beta$ are deleted and we apply the rule for σ_i, i.e. a rule of the form $\sigma_i \rightarrow A_i$ (here $A_i \in \{\Sigma \cup H\}^*$), by appending the word A_i onto the right end of W. So a computation step of a tag system is deterministic and is given as follows:

$$\sigma_i w_2 w_3 \ldots w_n \quad \vdash \quad w_{\beta+1} \ldots w_n A_i \tag{2}$$

A tag system halts if $|W| < \beta$ or if the leftmost symbol in the dataword is a halt symbol (i.e. $w_1 \in H$). It is not difficult to see that for each arbitrary tag system $\mathcal{T}$ there is an equivalent generalized tag system of the form given in Table 3. We say that the two systems are equivalent as the sequence of datawords produced by $\mathcal{T}$ is the same as the sequence of datawords produced by the generalized tag system at times t where $0 = t \bmod \beta$. That is given a word W as input, $\mathcal{T}$ produces the dataword W' in n time steps if and only if the generalized tag system in Table 3 also produces dataword W' in βn time steps when given W as input. To see this note that β steps of the generalized tag system in Table 3 on a configuration of the form $(0, \sigma_i w_2 w_3 \ldots w_n)$ gives

$$(0, \sigma_i w_2 w_3 \ldots w_n) \quad \vdash^\beta \quad (0, w_{\beta+1} \ldots w_n A_i)$$

and so produces the same dataword as the arbitrary tag system computation step in Eq. (2).

Table 3. Arbitrary tag system $\mathcal{T}$ with alphabet $\Sigma = \{\sigma_1, \sigma_2, \ldots, \sigma_m\}$, rules of the form $\sigma_i \rightarrow A_i$ and deletion number β converted to its equivalent generalized tag system

k	Σ				
	σ_1	σ_2	σ_3	$\cdots$	σ_m
0	A_1	A_2	A_3	$\cdots$	A_m
1	ϵ	ϵ	ϵ	$\cdots$	ϵ
$\vdots$	$\vdots$	$\vdots$	$\vdots$	$\cdots$	$\vdots$
$\beta - 1$	ϵ	ϵ	ϵ	$\cdots$	ϵ

2.2.2 Cyclic Tag Systems

Definition 3. *A cyclic tag system $\mathcal{C} = A_0, A_1, \ldots, A_{\beta-1}$ is a list of words $A_i \in \{0,1\}^*$ called appendants.*

A *configuration* of a cyclic tag system is a pair (k, W) where k is a number that points to an appendant A_k and $W = w_1 w_2 \ldots w_n$ is a binary word (where each $w_j \in \{0,1\}$). A computation step acts on a configuration by deleting w_1, incrementing k to $((k+1) \bmod \beta)$, and appending the word A_k onto the right end of W if $w_1 = 1$ and appending nothing to W if $w_1 = 0$. So the two possible cases for a cyclic tag system *computation step* are given by:

$$(k, 0w_2 \ldots w_n) \quad \vdash \quad ((k+1) \bmod \beta, w_2 \ldots w_n)$$
$$(k, 1w_2 \ldots w_n) \quad \vdash \quad ((k+1) \bmod \beta, w_2 \ldots w_n A_k)$$

Intuitively the list $\mathcal{C}$ is a program with k pointing to instruction A_k. In the initial configuration $k = 0$ and W is the binary input word. A cyclic tag system completes its computation if (i) the dataword is the empty word or (ii) it enters a repeating sequence of configurations. It is clear that a cyclic tag system is an instance of a generalized tag system of the form given in Table 4. The cyclic tag system given in Table 5 computes iterations of the Collatz function: given a configuration of the form $(0, (10)^n)$ it computes $(0, (10)^{f(n)})$ in two rounds over the dataword. Below is an example of the system in Table 5 started on $(0, (10)^3)$ computing $f(3) = 5$ followed by $f(5) = 8$ and $f(8) = 4$.

$$(0, (10)^3) \quad \vdash^6 \quad (2, (0100)^2) \quad \vdash^8 \quad (2, (10)^6) \quad \vdash^2 \quad (0, (10)^5)$$
$$\vdash^{10} \quad (2, (0100)^3) \quad \vdash^{12} \quad (2, (10)^9) \quad \vdash^2 \quad (0, (10)^8) \quad \vdash^{32} \quad (0, (10)^4)$$

Table 4. A cyclic tag system is a generalized tag system with a binary alphabet where all the appendants for one of the symbols are the empty word

k	Σ	
	1	0
0	$A_{0,1}$	ϵ
1	$A_{1,1}$	ϵ
$\vdots$	$\vdots$	$\vdots$
$\beta - 1$	$A_{\beta-1,1}$	ϵ

Table 5. A cyclic tag system that simulates iterations of the Collatz function

k	Σ	
	1	0
0	0100	ϵ
1	10	ϵ
2	ϵ	ϵ
3	101010	ϵ

For an arbitrary value x the system in Table 5 computes an iteration of the Collatz function as follows:

$$(0, (10)^x) \quad \vdash^{2x} \quad (0, (0100)^{\frac{x}{2}}) \quad \vdash^{2x} \quad (0, (10)^{\frac{x}{2}})$$

$$\text{if} \quad 0 = x \bmod 2$$

$$(0, (10)^x) \quad \vdash^{2x} \quad (2, (0100)^{\frac{x+1}{2}}) \vdash^{2x+2} (2, (10)^{\frac{3x+3}{2}}) \quad \vdash^{2} \quad (0, (10)^{\frac{3x+1}{2}})$$

$$\text{if} \quad 1 = x \bmod 2$$

During the first pass for each $(10)^2$ we cyclic through the program once appending a single 0100 thus producing $(0100)^{\frac{x}{2}}$, and if x is odd the final 10 pair appends an 0100 without completing the final cycle giving $(0100)^{\frac{x+1}{2}}$. If x is even we begin the second pass at program row 0 and so each 0100 that is read appends 10 giving $(10)^{\frac{x}{2}}$. If x is odd we begin the second pass at program row 2 and so each 0100 that is read appends $(10)^3$ giving $(10)^{\frac{3x+3}{2}}$ and after the last 0100 is read we take 2 further steps to move program control to row 0 by reading a 10 pair giving $(10)^{\frac{3x+1}{2}}$.

3   Simulating Generalized Tag Systems in Linear Time

In Sect. 2.2.1 we saw that tag systems are simulated by generalized tag systems in linear time without any encoding to the tag systems dataword. In Sect. 2.2.2 we noted that cyclic tag systems are simply a restricted form of generalized tag system. In this section we show that tag systems and cyclic tag system simulate generalized tag systems in linear time.

Theorem 1. *Given an arbitrary generalized tag system $\mathcal{G}$ with cycle number β there is a tag system $\mathcal{T}$ that simulates t steps of $\mathcal{G}$ in time $t + \lceil \frac{t}{\beta} \rceil$.*

Proof. We construct a tag system $\mathcal{T}$ with deletion number β that simulates the arbitrary generalized tag system given in Table 1. The alphabet Σ' of $\mathcal{T}$ is obtained from the alphabet Σ of $\mathcal{G}$ as follows: for each $\sigma_i \in \Sigma$ there are β symbols $\sigma_{1,i}, \sigma_{2,i}, \sigma_{3,i}, \ldots \sigma_{\beta-1,i} \in \Sigma'$. There is one further symbol e in Σ' and so we have

$$\Sigma' = \{e, \sigma_{0,1}, \sigma_{1,1}, \ldots \sigma_{\beta-1,1}, \sigma_{0,2}, \sigma_{1,2}, \ldots \sigma_{\beta-1,2}, \ldots \sigma_{0,m}, \sigma_{1,m}, \ldots \sigma_{\beta-1,m}\}.$$

The set of halt symbols H' of $\mathcal{T}$ is similarly obtained from $\mathcal{G}$'s set of halt symbols $H = \{h_1, h_2, \ldots h_s\}$ giving

$$H' = \{h_{0,1}, h_{1,1}, \ldots h_{\beta-1,1}, h_{0,2}, h_{1,2}, \ldots h_{\beta-1,2}, \ldots h_{0,s}, h_{1,s}, \ldots h_{\beta-1,s}\}.$$

An arbitrary configuration $(k, w_1 w_2 w_3 \ldots w_n)$ of $\mathcal{G}$ where each $w_j \in \Sigma \cup H$ is encoded as the tag system dataword

$$\langle w_1 \rangle \langle w_2 \rangle \langle w_3 \rangle \ldots \langle w_n \rangle \tag{3}$$
$$_{[k]}$$

where $\langle \sigma_i \rangle = e\sigma_{\beta-1,i}, \ldots \sigma_{1,i}, \sigma_{0,i}$ and $\langle h_i \rangle = eh_{\beta-1,i}, \ldots h_{1,i}h_{0,i}$, and $\langle w_1 \rangle$ is $[k]$
the word obtained by deleting all but the rightmost $k+1$ symbols from $\langle w_1 \rangle$ to give a word of the form $\sigma_{k,i}, \ldots \sigma_{1,i}, \sigma_{0,i}$ or $h_{k,i}, \ldots h_{1,i}h_{0,i}$.

In $\mathcal{T}$ the rule for e is $e \to \epsilon$ and all other rules have the form $\sigma_{k,i} \to \langle A_{k,i} \rangle$ where $(k, \sigma_i) \to A_{k,i}$ is a rule in $\mathcal{G}$, with $A_{k,i} = a_{k,i,1}a_{k,i,2} \ldots a_{k,i,l}$ and $a_{k,i,p} \in \Sigma \cup H$ for $1 \leqslant p \leqslant l$, and $\langle A_{k,i} \rangle = \langle a_{k,i,1} \rangle \langle a_{k,i,2} \rangle \ldots \langle a_{k,i,l} \rangle$.

Equation (4) below gives an arbitrary computation step of $\mathcal{G}$. Equation (5) shows how $\mathcal{T}$ simulates the step in Eq. (4) for the case $k < \beta - 1$. From (3) above the configurations on the left and right of Eq. (5) respectively encode the configurations on the left and right of Eq. (4) and so it only remains to verify that the dataword on the left of Eq. (5) produces the dataword on the right.

$$(k, \sigma_i w_2 \ldots w_n) \quad \vdash \quad ((k+1) \bmod \beta, w_2 \ldots w_n A_{k,i}) \qquad (4)$$

$$\underset{[k]}{\langle \sigma_i \rangle} \langle w_2 \rangle \langle w_3 \rangle \ldots \langle w_n \rangle \quad \vdash \quad \langle w_2 \rangle \langle w_3 \rangle \ldots \langle w_n \rangle \underset{[k+1]}{\langle A_{k,i} \rangle} \qquad (5)$$

$$\underset{[\beta-1]}{\langle \sigma_i \rangle} \langle w_2 \rangle \langle w_3 \rangle \ldots \langle w_n \rangle \quad \vdash^2 \quad \langle w_2 \rangle \langle w_3 \rangle \ldots \langle w_n \rangle \underset{[0]}{\langle A_{k,i} \rangle} \qquad (6)$$

From above $\underset{[k]}{\langle \sigma_i \rangle} = \sigma_{k,i}, \ldots \sigma_{1,i}, \sigma_{0,i}$ and so since $\sigma_{k,i}$ is the leftmost symbol in the dataword on the left of Eq. (5), $\mathcal{T}$ applies the rule $\sigma_{k,i} \to \langle A_{k,i} \rangle$ to append $\langle A_{k,i} \rangle$ as shown in the configuration on the right. Recall that the deletion number is β and so $\mathcal{T}$ deletes $\underset{[k]}{\langle \sigma_i \rangle}$ (which has length $k+1$) and the first $\beta - k - 1$ symbols of $\langle w_2 \rangle$. Since $|\langle w_2 \rangle| = \beta + 1$ deleting it leftmost $\beta - k - 1$ symbols gives $\underset{[k+1]}{\langle w_2 \rangle}$ as shown in the dataword on the right of Eq. (5). We have now shown that given the dataword on the left of Eq. (5) $\mathcal{T}$ takes a single step to produces the dataword on the right of Eq. (5). Equation (6) covers the remaining case of the step in Eq. (4) when $k = \beta - 1$. In this case $\mathcal{T}$ takes 2 computation steps as shown in Eq. (6) the first step is similar to the step taken in Eq. (5) it appends $\langle A_{k,i} \rangle$ and deletes $\underset{[\beta-1]}{\langle \sigma_i \rangle}$ without deleting any symbols in $\langle w_2 \rangle$ since $|\underset{[\beta-1]}{\langle \sigma_i \rangle}| = \beta$. For the second step in Eq. (6) the leftmost symbol in $\langle w_2 \rangle$ is read, which we know from above must be e, and so we apply the rule $e \to \epsilon$ which appends the empty word and deletes the first β symbols from $\langle w_2 \rangle$ giving $\underset{[0]}{\langle w_2 \rangle}$. So given dataword on the left of Eq. (6) $\mathcal{T}$ takes a two step to produces the dataword on the right of Eq. (6).

When $k < \beta - 1$ $\mathcal{T}$ simulates a step of $\mathcal{G}$ in one step (Eq. (5)) and when $k = \beta - 1$ $\mathcal{T}$ simulates a step of $\mathcal{G}$ in two steps (Eq. (6)) and so $\mathcal{T}$ simulates t steps of $\mathcal{G}$ in time $t + \lceil \frac{t}{\beta} \rceil$.   □

Theorem 2. *Given an arbitrary generalized tag system $\mathcal{G}$ with alphabets Σ and H, there is a cyclic tag system $\mathcal{C}$ that simulates t steps of $\mathcal{G}$ in time tr, where $r = |\Sigma| + |H|$.*

Proof. We give a cyclic tag system $\mathcal{C}$ that simulates the arbitrary generalized tag system given in Table 1. This cyclic tag system is given by

$$\mathcal{C} = \langle A_{0,1}\rangle, \langle A_{0,2}\rangle, \ldots \langle A_{0,m}\rangle, \epsilon_1, \epsilon_2, \ldots \epsilon_{r-m},$$
$$\langle A_{1,1}\rangle, \langle A_{1,2}\rangle, \ldots \langle A_{1,m}\rangle, \epsilon_1, \epsilon_2, \ldots \epsilon_{r-m},$$
$$\vdots \qquad\qquad \vdots$$
$$\langle A_{\beta-1,1}\rangle\langle A_{\beta-1,2}\rangle, \ldots \langle A_{\beta-1,m}\rangle, \epsilon_1, \epsilon_2, \ldots \epsilon_{r-m},$$

where $r - m = |H|$, $|\Sigma| = m$, $A_{k,i} = a_{k,i,1}a_{k,i,2}\ldots a_{k,i,l}$ with $a_{k,i,p} \in \Sigma \cup H$ for $1 \leqslant p \leqslant l$, and $\langle A_{k,i}\rangle = \langle a_{k,i,1}\rangle\langle a_{k,i,2}\rangle \ldots \langle a_{k,i,l}\rangle$ with $\langle \sigma_i\rangle = 0^{i-1}10^{r-i}$ and $\langle h_j\rangle = 0^{m+j-1}10^{r-m-j}$.

An arbitrary configuration $(k, w_1w_2w_3 \ldots w_n)$ of $\mathcal{G}$ where each $w_j \in \Sigma \cup H$ is encoded as the cyclic tag system configuration

$$(rk, \langle w_1\rangle\langle w_2\rangle\langle w_3\rangle \ldots \langle w_n\rangle) \tag{7}$$

An arbitrary computation step of $\mathcal{G}$ is given in Eqs. (8), and (9) shows how $\mathcal{C}$ takes r steps to simulate this single step of $\mathcal{G}$. From (7) above the configurations on the left and right of Eq. (9) respectively encode the configurations on the on the left and right of Eq. (8), and so it only remains to verify that the dataword on the left of Eq. (9) produces the dataword on the right of Eq. (9). In the configuration on the left of Eq. (9) the value rk points to appendant $\langle A_{k,1}\rangle$ in $\mathcal{C}$'s program. The word $\langle \sigma_i\rangle = 0^{i-1}10^{r-i}$ is read in r steps as shown in Eq. (9). Cyclic tag systems appending nothing when reading a 0 and so the 0's in $\langle \sigma_i\rangle$ append nothing. The first i steps when we read 0^{i-1} the pointer is incremented by one at each step and so when we read the single 1 in $\langle \sigma_i\rangle$ the pointer is at $\langle A_{k,i}\rangle$ and so $\langle A_{k,i}\rangle$ gets appended.

$$(k, \sigma_i w_2 \ldots w_n) \quad \vdash \quad ((k+1) \bmod \beta, w_2 \ldots w_n A_{k,i}) \tag{8}$$
$$(rk, \langle \sigma_i\rangle\langle w_2\rangle \ldots \langle w_n\rangle) \quad \vdash^r \quad (r(k+1) \bmod \beta r, \langle w_2\rangle \ldots \langle w_n\rangle\langle A_{k,i}\rangle) \tag{9}$$

$\square$

4   Relative Prime Tag Systems and the Impossibility of Linear Time Simulation Using Simple Encodings

In Definition 4 below we define what we call a simple encoding. Note that in Theorems 1 and 2 we respectively proved that tag systems and cyclic tag systems simulate generalized tag systems in linear time using encodings that satisfy Definition 4.

Definition 4 (Simple encoding). *Given alphabets Σ and Σ' we call an encoding function $f : \Sigma^* \to \Sigma'^*$ simple if $f(w_1w_2\ldots w_n) = \langle w_1\rangle\langle w_2\rangle\ldots\langle w_n\rangle$ where $w_i \in \Sigma$, $\langle w_i\rangle \in \Sigma'^*$ and for all $w_i = w_j$ where $i \neq j$, $\langle w_i\rangle = \langle w_j\rangle$.*

Before we continue we introduce some further notation and technical terms for tag systems that will be used in Lemma 3. For a tag system we say a symbol σ_i is *read* if and only if at the start of a computation step it is the leftmost symbol (i.e. the rule $\sigma_i \to A_i$ is applied), and we say a word $W = w_1w_2\ldots w_n$ is *entered with shift* $z < \beta$ if w_{z+1} is the leftmost symbol that is read in W (see for example Fig. 1). We use the term *round* to describe the $\lceil\frac{|W|}{\beta}\rceil$ computation steps that traverse a word W exactly once. A word W has a *shift change* of $0 \leqslant s < \beta$ if $|w| = y\beta - s$ where $y \in \mathbb{N}$ and $y > 0$ (for example the word V in Fig. 1 has a shift change of 2). The proof of Lemma 1 is left to the reader.

Lemma 1 (shift change). *Given a tag system T with deletion number β and the word $UW \in \Sigma^*$, where the word U has a shift change of s and $|W| \geqslant \beta$, after one round of T on U entered with shift z the word W is entered with shift $(z + s) \bmod \beta$.*

Theorem 3. *Let T and T' be tag systems with deletion number β and β' respectively such that there exists $p \in \mathbb{N}$ with $0 = \beta \bmod p$ and $0 \neq \beta' \bmod p$, then using a simple encoding T' cannot simulate T in linear time.*

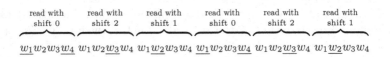

Fig. 1. Sequence of symbols read in dataword with deletion number 3. The symbols that will be read are marked with an underline. The entire dataword consists of the word $V = w_1w_2w_3w_4$ repeated 6 times. The leftmost occurrence of V is entered with shift 0 which means w_1 and w_4 are read, the second V from the left is entered with shift 2 which means w_3 is read, etc.

Proof. Let $\Sigma = \{\sigma_1, \sigma_2, \ldots \sigma_m\}$ be the alphabet of T. Below we show that if a simple encoding is used to encode $(\sigma_i)^s$, there exists no $c \in \mathbb{N}$ such that in c rounds (i.e. linear time) T' simulates T reading words of the form $(\sigma_i)^s$. From Definition 4 a simple encoding function gives a dataword of the form $f((\sigma_i)^s) = \langle\sigma_i\rangle^s$. The deletion number of T is β so it only reads $\lceil\frac{s}{\beta}\rceil$ symbols in $(\sigma_i)^s$. To simulate this T' must distinguish exactly $\lceil\frac{s}{\beta}\rceil$ of the $\langle\sigma_i\rangle$ words in $\langle\sigma_i\rangle^s$. Note that the only way a tag system can distinguish between identical words is to enter the words with different shift values so that different sequences of symbols are read in each word. In Fig. 1 the rules of the tag system can be used to record which shift value each V was entered with by mapping each word to a new word. For example if V is entered with shift 0, 1, or 2 it can be mapped to the words V_0, V_1 and V_2 respectively. So from Fig. 1 after one round over the word V^6 we would get the word $V_0V_1V_2V_0V_1V_2$. With another round we can further distinguish each word by mapping V_i ($i \in \{0,1,2\}$) to the words $V_{i,0}$, $V_{i,1}$ or $V_{i,2}$ when it is entered with shifts of 0, 1, or 2 respectively, and we can continue in this way with subsequent rounds over the dataword.

In Fig. 2 we show how the computation of $\mathcal{T}'$ progresses with each subsequent round over the dataword. The deletion number of $\mathcal{T}'$ is β' which means there are only β' possible shift values with which a word can be entered. So during the first round after reading $l_0 + 1$ of the $\langle \sigma_i \rangle$ words (where $l_0 \leqslant \beta'$), two $\langle \sigma_i \rangle$ words will have been entered with the same shift value. It follows that the sequence of shift values for the first l_0 $\langle \sigma_i \rangle$ words will be the same as the sequence of shift values for the second l_0 $\langle \sigma_i \rangle$ words and the third and so on as shown at the top of Fig. 2. Recall that the shift value determines which symbols are read in each word, and so the sequence of symbols read in each $\langle \sigma_i \rangle^{l_0}$ word entered with shift $x_{0,1}$ remains the same through out the dataword. So if we let W_1 be the word appended when the word $\langle \sigma_i \rangle^{l_0}$ is read with shift $x_{0,1}$, after one round we get a dataword of the form $W_1^{\frac{s}{l_0}}$ as shown in Fig. 2. The same arguments as those we have just made can be used to show that the subsequent datawords produced are of the form given in Fig. 2.

From above we know that a $\langle \sigma_i \rangle^{l_0}$ word produces a W_1 word in one round and a $(W_1)^{l_1}$ word produces a W_2 word in one round, and so a $\langle \sigma_i \rangle^{l_0 l_1}$ word produces W_2 in two rounds. Similarly a $\langle \sigma_i \rangle^{l_0 l_1 l_2}$ word produces a W_3 word in three rounds and a $\langle \sigma_i \rangle^{l_0 l_1 \dots l_{c-1}}$ word produces W_c in c rounds. It follows from Fig. 2 that if a pair of $\langle \sigma_i \rangle$ words are separated by $v l_0 l_1 \dots l_{c-1} - 1$ $\langle \sigma_i \rangle$ words (for $v \in \mathbb{N}$) in the initial dataword they both produce the same word after c rounds. To see this we write W_c as $W_c = W_{c,1} W_{c,2} \dots W_{c,l_0 l_1 \dots l_{c-1}}$ where in the word $\langle \sigma_i \rangle^{l_0 l_1 \dots l_{c-1}}$ that produces W_c the leftmost $\langle \sigma_i \rangle$ produces $W_{c,1}$, the second $\langle \sigma_i \rangle$ from the left produces $W_{c,2}$, the third produces $W_{c,3}$ and so on. So if a pair of $\langle \sigma_i \rangle$ words are separated by $v l_0 l_1 \dots l_{c-1} - 1$ $\langle \sigma_i \rangle$ words they both produce the word $W_{c,r}$ in W_c words that are separated by $v - 1$ W_c words in the final configuration of Fig. 2.

Recall that $\mathcal{T}'$ is attempting to distinguish $\lceil \frac{s}{\beta} \rceil$ of the $\langle \sigma_i \rangle$ words so it can simulate $\mathcal{T}$ reading $\lceil \frac{s}{\beta} \rceil$ of the σ_i symbols. Assuming that each $W_{c,j}$ word (for $1 \leqslant j \leqslant l_0 l_1 \dots l_{c-1}$) occurs only once in W_C, $\mathcal{T}'$ can distinguish $\lceil \frac{s}{l_0 l_1 \dots l_{c-1}} \rceil$ (or $\lfloor \frac{s}{l_0 l_1 \dots l_{c-1}} \rfloor$) of the $\langle \sigma_i \rangle$ words by choosing one of the $W_{c,j}$ words as there are $\lceil \frac{s}{l_0 l_1 \dots l_{c-1}} \rceil$ (or $\lfloor \frac{s}{l_0 l_1 \dots l_{c-1}} \rfloor$) of the W_c words and as mentioned above a single $\langle \sigma_i \rangle$ produces a single $W_{c,j}$ word. So the l_i values determine how many of the s $\langle \sigma_i \rangle$ values $\mathcal{T}'$ distinguishes. Below we show that the l_i values are relative prime to p a factor of β and so $\mathcal{T}'$ cannot distinguish $\lceil \frac{s}{\beta} \rceil$ of the $\langle \sigma_i \rangle$ words. Since $0 \neq \beta' \bmod p$ we can we prove that l_i is relative prime to p by showing that each l_i value is either 1, β' or a factor of β'. Recall that the value l_i tells us how many W_i words we must read before we have entered two W_i words with the same shift. From Lemma 1 if a word has a shift change of 0 the next word is entered with the same shift, which means that l_i must give a word $(W_i)^{l_i}$

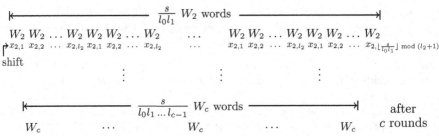

Fig. 2. Successive words produced by T' on each subsequent round of the dataword when the word $\langle \sigma_i \rangle^s$ is read on the first round. Each word produced by T' following y rounds on the dataword has the form $W_y{}^*$ where $W_y \in \Sigma'^*$. Below each W_y word is given the shift value $x_{y,j}$ with which the word is entered. Note that in each dataword above the rightmost W_y may not be complete.

that has a shift change of 0 (i.e. $0 = l_i |W_i| \bmod \beta'$). If $0 = |W_i| \bmod \beta'$ then $l_i = 1$. If $gcd(|W_i|, \beta') = 1$ then $l_i = \beta'$ is the smallest natural number that satisfies $0 = l_i |W_i| \bmod \beta'$. If $gcd(|W_i|, \beta') = r$ with $1 < r < \beta'$ then $l_i = \frac{\beta'}{r}$ is the smallest natural number that satisfies $0 = l_i |W_i| \bmod \beta'$. So $l_0 l_1 \ldots l_{c-1}$ is a product of 1, β', and factors of β' and so does not have p as a factor. From above there are at most $l_0 l_1 \ldots l_{c-1}$ different types of $W_{c,j}$ word with each $W_{c,j}$ word allowing T' to distinguish either $\lceil \frac{s}{l_0 l_1 \ldots l_{c-1}} \rceil$ (or $\lfloor \frac{s}{l_0 l_1 \ldots l_{c-1}} \rfloor$) encoded σ_i symbols and so T' may choose $z \leqslant l_0 l_1 \ldots l_{c-1}$ different types of $W_{c,j}$ word to distinguish between $z \lceil \frac{s}{l_0 l_1 \ldots l_{c-1}} \rceil$ and $z \lfloor \frac{s}{l_0 l_1 \ldots l_{c-1}} \rfloor$ encoded σ_i symbols. However for $z \lceil \frac{s}{l_0 l_1 \ldots l_{c-1}} \rceil \leqslant r \leqslant z \lfloor \frac{s}{l_0 l_1 \ldots l_{c-1}} \rfloor$ we know that $r \neq \lceil \frac{s}{\beta} \rceil$ for sufficiently large s values as p is a factor of β and we have shown that p is not a factor of $l_0 l_1 \ldots l_{c-1}$. It follows that when using a simple encoding, T' cannot distinguish the correct number of encoded read symbols to give a linear time simulation of a round of T on words of the form $(\sigma_i)^s$.   □

Corollary 1. *Let T be a tag system with deletion number β and $\mathcal{G}'$ be a generalized tag system with cycle number β' such that there exists $p \in \mathbb{N}$ with $0 = \beta \bmod p$ and $0 \neq \beta' \bmod p$, then using a simple encoding $\mathcal{G}'$ cannot simulate T in linear time.*

Proof. In the proof of Theorem 1 we give a tag system with deletion number β that simulates a generalized tag system that has cycle number β. The simulation runs in linear time and uses a simple encoding (Definition 4). If $\mathcal{G}'$ uses a simple encoding to simulate T in linear time, then we can give a tag system $T_{\mathcal{G}}$ that simulates T in linear time and uses a simple encoding but has a deletion number of β'. It follows from Theorem 3 that such a simulation is not possible.   □

References

1. Baiocchi, C.: 3n+1, UTM e tag-system. Technical report Pubblicazione 98/38, Dipartimento di Matematico, Università di Roma (1998). (In Italian)
2. Baiocchi, C., Margenstern, M.: Cellular automata about the 3x+1 problem. In: Proceedings of LCCS 2001, Université Paris 12, pp. 37–45 (2001)
3. Cocke, J., Minsky, M.: Universality of tag systems with $P = 2$. J. ACM **11**(1), 15–20 (1964)
4. Cook, M.: Universality in elementary cellular automata. Complex Syst. **15**(1), 1–40 (2004)
5. Cook, S.: The solvability of the derivability problem for one-normal systems. J. ACM **13**(2), 223–225 (1966)
6. De Mol, L.: Tag systems and Collatz-like functions. Theoret. Comput. Sci. **390**(1), 92–101 (2008)
7. De Mol, L.: Solvability of the halting and reachability problem for binary 2-tag systems. Fundamenta Informaticae **99**(4), 435–471 (2010)
8. Lagarias, J.C.: The Ultimate Challenge: The 3x+1 Problem. American Mathematical Society, Providence (2010)
9. Lagarias, J.C.: The 3x+1 problem: an annotated bibliography (1963–1999). Technical report arXiv:math/0309224v13 [math.NT], January 2011
10. Lagarias, J.C.: The 3x+1 problem: an annotated bibliography, ii (2000–2009). Technical report arXiv:math/0608208v6 [math.NT], February 2012
11. Margenstern, M.: Frontier between decidability and undecidability: a survey. Theoret. Comput. Sci. **231**(2), 217–251 (2000)
12. Michel, P.: Busy beaver competition and Collatz-like problems. Archive Math. Logic **32**(5), 351–367 (1993)
13. Michel, P.: Small Turing machines and the generalized busy beaver competition. Theoret. Comput. Sci. **326**, 45–56 (2004)
14. Neary, T.: Small universal Turing machines. Ph.D. thesis, Department of Computer Science, National University of Ireland, Maynooth (2008)
15. Neary, T.: Undecidability in binary tag systems and the Post correspondence problem for five pairs of words. In: 32nd International Symposium on Theoretical Aspects of Computer Science, STACS. LIPIcs, vol. 30, pp. 649–661 (2015)
16. Neary, T., Woods, D.: P-completeness of cellular automaton Rule 110. In: Bugliesi, M., Preneel, B., Sassone, V., Wegener, I. (eds.) ICALP 2006. LNCS, vol. 4051, pp. 132–143. Springer, Heidelberg (2006). https://doi.org/10.1007/11786986_13

17. Post, E.: Formal reductions of the general combinatorial decision problem. Am. J. Math. **65**(2), 197–215 (1943)
18. Wang, H.: Tag systems and lag systems. Math. Ann. **152**(4), 65–74 (1963)
19. Woods, D., Neary, T.: On the time complexity of 2-tag systems and small universal Turing machines. In: 47th Annual IEEE Symposium on Foundations of Computer Science (FOCS), pp. 439–448, October 2006

Certain Query Answering on Compressed String Patterns: From Streams to Hyperstreams

Iovka Boneva[1], Joachim Niehren[2], and Momar Sakho[2(✉)]

[1] Université de Lille, Lille, France
[2] Inria Lille, Lille, France
momar.sakho@inria.fr

Abstract. We study the problem of certain query answering (CQA) on compressed string patterns. These are incomplete singleton context-free grammars, that can model systems of multiple streams with references to others, called hyperstreams more recently. In order to capture regular path queries on strings, we consider nondeterministic finite automata (NFAs) for query definition. It turns out that CQA for Boolean NFA queries is equivalent to regular string pattern inclusion, i.e., whether all strings completing a compressed string pattern belong to a regular language. We prove that CQA on compressed string patterns is PSPACE-complete for NFA queries. The PSPACE-hardness even applies to Boolean queries defined by deterministic finite automata (DFAs) and without compression. We also show that CQA on compressed linear string patterns can be solved in PTIME for DFA queries.

1 Introduction

A stream is a sequence of events that arrive incrementally one by one from the left to the right. Most typically, streams are produced by social networks such as Twitter, database systems as for producing financial transactions, information systems, sensor systems, or more generally when communicating semi-structured data over the internet. We are interested in the problem of monitoring streams in a reactive manner [16, 22, 23, 25]. The objective is to select the relevant events of a stream as quickly as possible upon their arrival. This requires to decide whether an event of the stream is a certain answer of the logical query that defines the relevant events of the monitoring task. Lowering the latency of this decision process increases the reactivity of the stream processing system and reduces its memory costs. A limitation to constant memory may seem ideal in theory, but is too restrictive for many monitoring tasks in practice. A less restrictive objective is thus to minimize the latency and thereby to reduce the memory consumption.

In the present paper we study a generalization of streams to multiple streams with references as introduced by Maneth, Ordóñez and Seidl [21]. The references point to unknown parts in the middle of a stream. The same reference may be

© Springer Nature Switzerland AG 2018
I. Potapov and P.-A. Reynier (Eds.): RP 2018, LNCS 11123, pp. 117–132, 2018.
https://doi.org/10.1007/978-3-030-00250-3_9

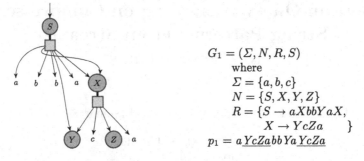

$$G_1 = (\Sigma, N, R, S)$$
$$\text{where}$$
$$\Sigma = \{a, b, c\}$$
$$N = \{S, X, Y, Z\}$$
$$R = \{S \rightarrow aXbbYaX,$$
$$X \rightarrow YcZa \quad\}$$
$$p_1 = a\underline{YcZa}bbYa\underline{YcZa}$$

Fig. 1. The hyperstream G_1 and its string pattern p_1.

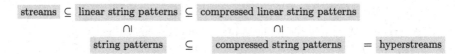

Fig. 2. Landscape from streams to hyperstreams.

used multiple times, allowing to share unknown parts. Streams with similar references were named hyperstreams in own previous work [20]. Here, we propose to formalize hyperstream containing words (rather than linearizations of trees or nested words) as *incomplete* versions of singleton context-free grammars [24] (also termed straight line programs [3]), where the rules of some nonterminals may be missing. The hyperstream $G_1 = (\Sigma, N, R, S)$ is illustrated graphically in Fig. 1. It has the terminals in $\Sigma = \{a, b, c\}$, the nonterminals in $N = \{S, X, Y, Z\}$, the set R with the rules $S \rightarrow aXbbYaX$ and $X \rightarrow YcZa$, and the start symbol S. The nonterminals are called the references of the hyperstream. For some of these references there exists a rule in the grammar, and if so, this rule is unique. For any grammar rule, the reference on its left is said to refer to the string pattern on its right. The hyperstream G_1 has a rule for S and X, while it misses those of Y and Z. The missing rules for these references may be added in the future one by one by the hyperstream's environment. Alternatively, hyperstreams can be identified with *compressed string patterns*. The hyperstream G_1 for instance represents the string pattern $p_1 = a\underline{YcZa}bbYa\underline{YcZa}$, while sharing the underlined factors substituted for the two occurrences of X. Streams are a special case of string patterns that have a unique occurrence of a variable in their last position. The landscape from streams to hyperstreams, over linear string patterns, string patterns, and compressed string patterns is illustrated in Fig. 2.

In this paper, we study the decision problem of certain query answering (CQA) on compressed string patterns, i.e., whether a tuple of positions is a certain answer of a query on a compressed string pattern. Here we consider the positions of the string pattern after decompression rather than the positions of the grammar. Intuitively, a tuple of positions is a certain query answer on a compressed string pattern G if it is an answer to the query on all completions of G, up to the offsets raised by the completion of G on its decompression. We will also consider the symmetric problem for certain query non-answers.

Motivated by regular path queries [1], we consider nondeterministic finite automata (NFAs) for defining such queries. For instance, the query $\mathbf{Q}_1$ on strings over $\Sigma = \{a, b, c\}$ that selects all a-positions that are followed eventually by bb can be defined by the following regular path in XPATH-like notation:

$$\mathbf{Q}_1 = \texttt{successor}^* :: a[\texttt{successor}^* :: b/\texttt{successor} :: b].$$

It can also be defined by the x-pointed regular expression $\Sigma^* a^x \Sigma^* bb\Sigma^*$ where x is a variable for the position that is to be selected. Now, consider the case, where the string is not given explicitly but only described partially by some (compressed) string pattern. On the string pattern p_1, for instance, the a-positions 1 and 5 are certain query answers for $\mathbf{Q}_1$, while the a-positions 9 and 13 are not. The position 13 is even a certain non-answer.

When restricted to Boolean NFA queries, CQA becomes equivalent to the problem of whether all strings described by the completions of a compressed string pattern are accepted by the NFA. For the string pattern Y (for some variable Y), this problem clearly generalizes on the universality problem of NFAs, which is well-known to be PSPACE-complete [17]. The following questions, however, are open to the best of our knowledge, even in the case of string patterns without compression: Is CQA on (compressed) string patterns decidable for NFA-defined queries, and if yes, what is the complexity? Does CQA on (compressed) string patterns remain hard for queries defined by deterministic finite automata (DFAs)? For which restrictions of (compressed) string patterns is CQA in PTIME? And what about the symmetric questions concerning certain query non-answers? The objective of the present paper is to answer these questions in all possible cases.

Our first contribution is that CQA on string patterns is PSPACE-complete, both for NFA queries and DFA queries, with and without compression, Boolean or not, see Fig. 3. This upper bound is not fully obvious, as the set of strings defined by a string pattern may be non-regular and even non-context-free. Furthermore, the lower bound may be surprising in that CQA for DFA queries on string patterns is more complex than on streams, where it is in PTIME (Theorem 1 of [13]), and also more complex than string pattern matching, which is NP-complete (Theorem 3.6 of [2]) even with compression (Theorem 4.10 of [11]).

Our second contribution is that CQA for DFA queries can be decided in PTIME on compressed *linear* string patterns, see Fig. 4. The linearity restriction matches with the worst case complexity for streams, even though linear compressed string patterns allow for unknown factors and compression in addition. This result (Corollary 2) is based on a novel algorithm for partial decompression of compressed string patterns that we present (Lemma 6), followed by a test of a reachability property (Theorem 3).

Our third contribution is that the certainty of query non-answers on compressed string patterns is PSPACE-complete, both for Boolean and non-Boolean queries, and independently of whether they are defined by DFAs or NFAs. In the

	DFAS	NFAS
Answers	PSPACE-c	PSPACE-c
Non-answers	PSPACE-c	PSPACE-c

	DFAS	NFAS
Answers	PTIME	PSPACE-c
Non-answers	PTIME	PTIME

Fig. 3. Query certainty on (compressed) string patterns or hyperstreams.

Fig. 4. Query certainty on (compressed) linear string patterns or streams.

Boolean case, the problem is equivalent to whether a compressed string pattern does *not* match the regular language accepted by the automaton. This problem generalizes on the complement of compressed string pattern matching, and thus is CONP-hard. So while certain query non-answering can be solved in PTIME on streams, the complexity increases to PSPACE on compressed string patterns. Finally, we show that the restriction of the problem to compressed linear string patterns – that is, *regular* compressed linear string pattern matching – can also be solved in PTIME even for queries defined by NFAS.

Outline. In Sect. 2 we start with some preliminaries on finite automata theory. Section 3 recalls the notion of compressed string patterns and in Sect. 4 we study the problems of regular compressed pattern inclusion and matching. Section 5 recalls how to define non-Boolean queries on strings by automata. In Sects. 6 and 7 we generalize the notions of certain query answers and non-answers to (compressed) string patterns and study their complexity. All omitted proofs can be found in the long online version[1].

Related Work. The notion of certain query answers for incomplete relational structures is standard in databases research [9]. In the context of stream processing, certain query answers were called answers that are safe for selection and certain query non-answers were called safe for rejection [12]. Certain query non-answers were studied for fast failure [4] and for reducing the memory consumption of streaming systems. The problem of certain query answering and non-answering on streams has been shown to be computationally hard even for queries defined in tiny fragments of first-order logic [12]. Certain query non-answering was shown to be hard in the context of online verification [4,19].

As shown by [12], those classes of queries on strings for which the problem of certain query answering on streams is known to be feasible, are either such that certainty is always determined with 0-delay [4,14,22]) or such that the queries in the class can be compiled to DFAS in PTIME.

Algorithms for processing XML streams or complex event streams raised much interest in the literature [15,16,25] and motivated the work on hyperstreams. XML streams contain nested words [18,22] rather than strings without bracket structure. The best existing algorithms for answering navigational XPATH queries (i.e. first-order logic queries) on XML streams are based on

---

[1] The long version can be found at the address https://hal.inria.fr/hal-01846016.

compilation to nested word automata [10,23]. Low but not lowest latency is achieved with high efficiency by approximating certain answers for queries defined by nondeterministic nested word automata.

2 Preliminaries

Words. The set of natural numbers with 0 is denoted by $\mathbb{N}$. For any set Σ, a word over Σ is a tuple $(a_1, \ldots, a_n) \in \Sigma^n$ where $n \in \mathbb{N}$. We denote such words by $a_1 \ldots a_n$ and by ϵ if $n = 0$. We denote the ith letter of a word $u = a_1 \ldots a_n$ by $u[i] =_{def} a_i$. The set of all words over Σ is denoted by Σ^*. The concatenation of two words $u_1, u_2 \in \Sigma^*$ is denoted by $u_1 \cdot u_2 \in \Sigma^*$. For instance, if $\Sigma = \{a, b\}$ then $aba \cdot a = abaa$. The set of positions of a word $u = a_1 \ldots a_n$ is $pos(u) = \{1, \ldots, n\}$. For any subset $\Sigma' \subseteq \Sigma$ the set $pos_{\Sigma'}(u)$ is the subset of positions i of u such that $a_i \in \Sigma'$. Given a word $w = a_1 \ldots a_n$ and a second word $u = b_1 \ldots b_n$ of the same length possibly on a different alphabet, we define the zipped word over the product alphabet by $w * u = (a_1, b_1) \ldots (a_n, b_n)$. As a convention throughout the paper, we use the term *string* for words over the default alphabet Σ, as opposed to words over other alphabets such as $\Sigma \cup \mathcal{Y}$ for string patterns, and $\Sigma^{\mathcal{V}}$ for $\mathcal{V}$-annotated strings, that will be introduced later on.

Monoids. A monoid is a triple $(M, \cdot^M, 1^M)$ where M is a set, $\cdot^M : M \times M \to M$ is an associative binary operation and $1^M \in M$ is the neutral element, i.e. $1^M \cdot^M m = m \cdot^M 1^M = m$ for all $m \in M$. Given a word $u = m_1 \ldots m_n \in M^*$ we define its evaluation by $u^M = m_1 \cdot^M \ldots \cdot^M m_n$ where $\epsilon^M = 1^M$.

Most typically, we will consider the monoid of words $(\Sigma^*, \cdot, \epsilon)$ on some alphabet Σ, with the concatenation operation $\cdot : \Sigma^* \times \Sigma^* \to \Sigma^*$, and the empty word ϵ as neutral element. Alternatively, given another set Q, we will consider the *transition monoid* $(T_Q, \circ, id)$, where $T_Q = 2^{Q \times Q}$ is the set of binary relations over Q, $\circ : T_Q \times T_Q \to T_Q$ is the composition operation of binary relations on Q, and $id = \{(q, q) \mid q \in Q\}$ is the identity transition.

Finite Automata. A nondeterministic finite-state automaton (NFA) is a tuple $A = (Q, \Sigma, \delta, I, F)$ where Q and Σ are finite sets, $\delta \subseteq Q \times \Sigma \times Q$, and $I, F \subseteq Q$. We call Q the state set, Σ the alphabet, δ the transition relation, I the set of initial states, and F the set of final states of the automaton. An NFA A is deterministic, or a DFA, if it has exactly one initial state and the transition relation δ forms a partial function from $Q \times \Sigma$ to Q. The elements of T_Q are called the *transitions* of A. Any transition relation $\delta : Q \times \Sigma \times Q$ can be extended homomorphically to a transition function $\delta : \Sigma^* \to T_Q$ that assigns to any string a transition of A. Here we overload the symbol δ to stand for the transition relation and the transition function. The transition $\delta(a)$ of a letter $a \in \Sigma$ is $\{(q, q') \mid (q, a, q') \in \delta\}$ and the transition $\delta(w)$ of a string $w = a_1 \ldots a_n \in \Sigma^*$ is $\delta(w) = (\delta(a_1) \ldots \delta(a_n))^{T_Q}$. This is the composition of the transitions of all letters of w based on the operations of the transition monoid T_Q, and its neutral

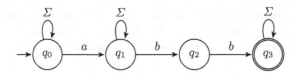

Fig. 5. Automaton A_2 defining the Boolean query $\mathbf{Q}_2$ with alphabet $\Sigma = \{a, b, c\}$

element for the empty word. Note that if A is a DFA then all transitions $\delta(w)$ are partial functions. A transition τ is called I, F-*successful* if $\tau \cap (I \times F) \neq \emptyset$. The language of an NFA $A = (Q, \Sigma, \delta, I, F)$ is the set $\mathcal{L}(A) = \{w \in \Sigma^* \mid \delta(w) \text{ is } I, F\text{-successful}\}$. The size of an automaton is $|A| = |Q| + |\delta|$.

Example 1. NFAs can be used to define Boolean queries on strings such as query $\mathbf{Q}_2$ which tests whether some position will be selected by query $\mathbf{Q}_1$, i.e., whether a string contains an a-letter followed eventually by a factor bb. The language of all such strings is defined by the automaton on Fig. 5.

A transition τ is called δ-*inhabited* if there exists a word $w \in \Sigma^*$ such that $\delta(w) = \tau$. *Transition inhabitation* Inh_Σ is the decision problem that receives as input a finite set Q, a transition relation $\delta \subseteq Q \times \Sigma \times Q$ and a transition $\tau \in T_Q$, and outputs whether τ is δ-inhabited. Inh_Σ is also called the *membership problem of the transition monoid* $\delta(\Sigma^*) \subseteq T_Q$ [7].

Theorem 1 (Kozen [17]). *For any set Σ with at least 2 elements, the transition inhabitation problem Inh_Σ is* PSPACE*-complete.*

The PSPACE hardness proof can be done by reduction from the problem of *non-emptiness of the intersection of sequences of* DFAs, which was shown to be PSPACE-complete in [17] too.

3   Compressed String Patterns

We fix an infinite set $\mathcal{Y}$ of *string variables* for the rest of the paper. A *string pattern* over a finite alphabet Σ is a word in $(\Sigma \cup \mathcal{Y})^*$. The set of all string patterns over Σ is denoted by Pat_Σ. The set of variables that occur in a string pattern p is denoted by $fv(p)$. An instance of a string pattern $p \in Pat_\Sigma$ is a string that can be obtained by substituting the variables of p by strings in Σ^*. Any substitution $\sigma : \mathcal{Y} \to \Sigma^*$ can be lifted to a substitution on string patterns $\hat{\sigma} : Pat_\Sigma \to \Sigma^*$ such that for all $p, p' \in Pat_\Sigma$, $a \in \Sigma$, and $Y \in \mathcal{Y}$: $\hat{\sigma}(pp') = \hat{\sigma}(p) \cdot \hat{\sigma}(p')$, $\hat{\sigma}(\epsilon) = \epsilon$, $\hat{\sigma}(a) = a$, and $\hat{\sigma}(Y) = \sigma(Y)$. We define the set of *instances* of a string pattern $p \in Pat_\Sigma$ as:

$$Inst(p) = \{\hat{\sigma}(p) \mid \sigma : \mathcal{Y} \to \Sigma^*\}.$$

For example, the string $ac\underline{cb}cbabb\underline{cb}a$ is an instance of the pattern $aYcZabbYa$, obtained with the substitution $[Y/cb, Z/b]$. A string pattern is called *linear*, if all its variables occur at most once. The set of all linear string patterns over Σ is denoted $LinPat_\Sigma$.

Definition 1 (Compressed string pattern). *A compressed string pattern is an* acyclic *CFG* $G = (N, \Sigma, R, S)$ *where* $N \subseteq \mathcal{Y}$ *is a finite set of nonterminals,* Σ *is an alphabet of terminal symbols disjoint from* $\mathcal{Y}$*, the ruling function* R *is a partial function that maps some of the nonterminals in* N *to string patterns in* $(N \cup \Sigma)^*$*, and* $S \in N$ *is the start symbol. The set of all compressed string patterns over* Σ *is denoted by* $cPat_\Sigma$*.*

We recall that a CFG G is acyclic if the binary relation $>_G = \{(Y, Z) \mid Y \in dom(R), Z \in fv(R(Y))\}$ is acyclic. The compressed string pattern from the introduction for instance has the rules $R(S) = aXbbYaX, R(X) = YcZa$. These rules induce the binary relation $\{(S, X), (S, Y), (X, Y), (X, Z)\}$ which is acyclic. The size of G is $|G| = |N| + \sum_{Y \in dom(R)} |R(Y)|$. The set of free variables of G is $fv(G) = N \setminus dom(R)$. A compressed string pattern G is called a *singleton context-free grammar* (sCFG) if it has no free variables, that is $fv(G) = \emptyset$. It is well-known that any sCFG defines a single string in Σ^*. The object of interest here is the set of strings that can be obtained by completing a compressed string pattern G to a sCFG, or equivalently, the set of instances of the string pattern of G defined as follows. For any compressed string pattern $G = (N, \Sigma, R, S)$, the grammar $G' = (N \setminus fv(G), \Sigma \cup fv(G), R, S)$ is a sCFG. We define the string pattern $pat(G) \in Pat_{\Sigma \cup fv(G)}$ as the unique word in the language of G'. Formally, for every $Y \in dom(R)$, let G_Y be the compressed string pattern $G_Y = (N, \Sigma, R, Y)$. If $S \in dom(R)$ then $pat(G) = \hat{\sigma}(R(S))$ where $\sigma(Y) = pat(G_Y)$ for all $Y \in dom(R) \setminus \{S\}$. This recursive definition is well-founded because G is an acyclic CFG. Otherwise, if $S \in fv(G)$, then $pat(G) = S$. For instance, if G_1 is the hyperstream from the introduction, then $pat(G_1) = p_1$.

A compressed string pattern G is called a *compressed linear string pattern* if $pat(G)$ is linear. The set of all compressed linear string patterns over Σ is denoted $cLinPat_\Sigma$. Finally for any string pattern $p \in Pat_\Sigma$ there exists a compressed string pattern G_p having p as pattern, namely $G_p = (\{S\} \cup fv(p), \Sigma, R, S)$ with $dom(R) = \{S\}$ and $R(S) = p$. Clearly $pat(G_p) = p$. Therefore we will identify p with G_p, so that $Pat_\Sigma \subseteq cPat_\Sigma$.

4   Regular Pattern Inclusion and Matching

We consider the problems of *regular compressed pattern inclusion*, i.e. testing whether all strings described by a completion of a compressed string pattern to a sCFG are accepted by a finite automaton, and of *regular compressed pattern matching*, whether some string described by a completion of a compressed string pattern is accepted by a finite automaton.

A class of compressed string patterns $\mathcal{G}$ is a function from finite sets Σ to subsets $\mathcal{G}_\Sigma \subseteq cPat_\Sigma$ such as for instance $\mathcal{G} \in \{Pat, cPat, LinPat, cLinPat\}$. A class of NFAs $\mathcal{A}$ is a function from finite sets Σ to subsets $\mathcal{A}_\Sigma \subseteq NFA_\Sigma$, where NFA_Σ is the set of NFAs with alphabet Σ. Most typically, $\mathcal{A} \in \{DFA, NFA\}$. For any class $\mathcal{G}$ of compressed string patterns, any class $\mathcal{A}$ of NFAs, and any finite set Σ we define the following two decision problems.

Regular compressed pattern inclusion. $\mathrm{INCL}_\Sigma(\mathcal{G}, \mathcal{A})$. *Input:* A compressed string pattern $G \in \mathcal{G}_\Sigma$ and a finite automaton $A \in \mathcal{A}_\Sigma$.
Output: The truth value of whether $Inst(pat(G)) \subseteq \mathcal{L}(A)$.

Regular compressed pattern matching. $\mathrm{MATCH}_\Sigma(\mathcal{G}, \mathcal{A})$. *Input:* A compressed string pattern $G \in \mathcal{G}_\Sigma$ and a finite automaton $A \in \mathcal{A}_\Sigma$.
Output: The truth value of whether $Inst(pat(G)) \cap \mathcal{L}(A) \neq \emptyset$.

The problem $\mathrm{COMATCH}_\Sigma(\mathcal{G}, \mathcal{A})$ is the complement of the problem MATCH_Σ $(\mathcal{G}, \mathcal{A})$, and thus outputs for a given compressed string pattern $G \in \mathcal{G}_\Sigma$ and a finite automaton $A \in \mathcal{A}_\Sigma$ whether $Inst(pat(G)) \cap \mathcal{L}(A) = \emptyset$.

Example 2. Any instance of $pat(G_1) = aYcZabbYaYcZa$ answers the Boolean query $\mathbf{Q}_2 = [\mathtt{successor}^* :: a/\mathtt{successor}^* :: b/\mathtt{successor} :: b]$ from Example 1, i.e., the instance set of $pat(G_1)$ is included in the language of NFA A_2 in Fig. 5.

Let $s\mathrm{DFA}$ be the subclass of DFAs that recognize a singleton language. Note that the well-known problem of string pattern matching is $\mathrm{MATCH}_\Sigma(Pat, s\mathrm{DFA})$, and $\mathrm{MATCH}_\Sigma(cPat, s\mathrm{DFA})$ is its extension with compression. We recall from [11] that string pattern matching with and without compression respectively are NP-complete for all alphabets Σ with at least 2 letters, but in PTIME when restricted to linear string patterns even with compression.

Our first main contribution is the following complexity result for regular compressed pattern matching and inclusion (see Fig. 3).

Theorem 2 (Non-linear patterns). *For any* $\mathcal{G} \in \{Pat, cPat\}$ *and* $\mathcal{A} \in \{\mathrm{DFA}, \mathrm{NFA}\}$ *and for any finite alphabet* Σ *with at least 2 letters, the problems of regular compressed pattern inclusion* $\mathrm{INCL}_\Sigma(\mathcal{G}, \mathcal{A})$ *and matching* $\mathrm{MATCH}_\Sigma(\mathcal{G}, \mathcal{A})$ *are* PSPACE-*complete.*

This shows that these problems are decidable even though the instance sets of nonlinear patterns like $Inst(YYY)$ are neither regular nor context-free. The theorem also shows that regular pattern matching $\mathrm{MATCH}_\Sigma(Pat, \mathrm{DFA})$ is PSPACE-complete and thus harder than compressed string pattern matching $\mathrm{MATCH}_\Sigma(Pat, s\mathrm{DFA})$ which is NP-complete.

Proof. We will present a sequence of PSPACE reductions from Lemma 1 until Lemma 4 that imply the theorem when composed as in Fig. 6.

For two decision problems A and B, we write $A =_p B$ when A reduces polynomially to B and B reduces polynomially to A.

Lemma 1. $\mathrm{INCL}_\Sigma(LinPat, \mathrm{NFA})$ *is* PSPACE-*hard if* $|\Sigma| \geq 2$.

The proof is straightforward by reduction from the universality of NFAs.

We now show the PSPACE upper bound for $\mathrm{INCL}_\Sigma(cPat, \mathrm{NFA})$. For any transition relation $\delta \subseteq Q \times \Sigma \times Q$ and any substitution into the transition monoid $\sigma : \mathcal{Y} \to T_Q$, we define δ^σ to be the function that takes a string pattern in Pat_Σ as input, and returns a transition, such that for all $p, p' \in Pat_\Sigma$, $w \in \Sigma^*$, and $Y \in \mathcal{Y}$:

$$\delta^\sigma(w) = \delta(w), \qquad \delta^\sigma(y) = \sigma(y), \qquad \delta^\sigma(pp') = \delta^\sigma(p) \circ \delta^\sigma(p').$$

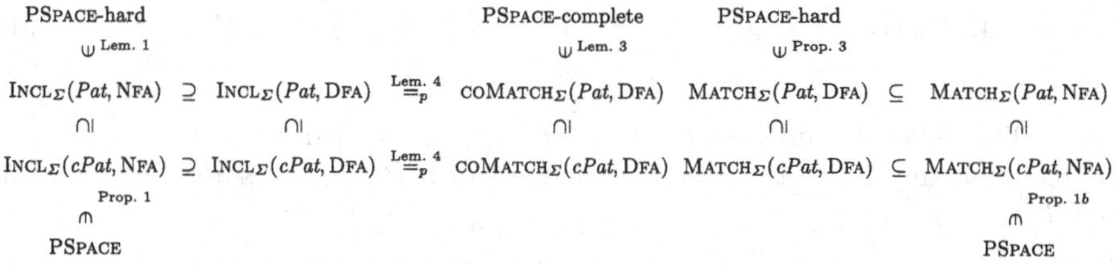

PSpace-hard   PSpace-complete   PSpace-hard
  ⊎ Lem. 1   ⊎ Lem. 3   ⊎ Prop. 3

INCL$_\Sigma$(*Pat*, NFA) ⊇ INCL$_\Sigma$(*Pat*, DFA) $\overset{\text{Lem. 4}}{=_p}$ coMATCH$_\Sigma$(*Pat*, DFA)   MATCH$_\Sigma$(*Pat*, DFA) ⊆ MATCH$_\Sigma$(*Pat*, NFA)

  ∩| ∩| ∩| ∩| ∩|

INCL$_\Sigma$(*cPat*, NFA) ⊇ INCL$_\Sigma$(*cPat*, DFA) $\overset{\text{Lem. 4}}{=_p}$ coMATCH$_\Sigma$(*cPat*, DFA)   MATCH$_\Sigma$(*cPat*, DFA) ⊆ MATCH$_\Sigma$(*cPat*, NFA)

  Prop. 1   Prop. 1b
  ⋔   ⋔
  PSpace   PSpace

Fig. 6. Regular inclusion and matching problems relationship and complexity classes

Lemma 2. *Given a transition relation* $\delta \subseteq Q \times \Sigma \times Q$ *of some* NFA*, a substitution* $\sigma : \mathcal{Y} \to T_Q$ *and a compressed string pattern* $G \in cPat_\Sigma$*, the transition* $\delta^\sigma(pat(G)) \in T_Q$ *can be computed in time* $O(|Q|^3|G|)$.

Proposition 1. INCL$_\Sigma$(*cPat*, NFA) *is in* PSpace.

Proof. Given an NFA A, a compressed string pattern G over Σ, we have to check whether $Inst(pat(G)) \subseteq \mathcal{L}(A)$. By definition, this is equivalent to checking whether $\delta(\hat{s}(pat(G)))$ is I, F-successful for every substitution $s : fv(G) \to \Sigma^*$. The latter is equivalent to checking whether $\delta^\sigma(pat(G))$ is I, F-successful for all substitution $\sigma : fv(G) \to T_Q$ that maps the free variables of G to δ-inhabited transitions. A decision procedure can thus enumerate all substitutions $\sigma : fv(G) \to T_Q$ to δ-inhabited transitions, compute $\delta^\sigma(pat(G))$ and check whether it is I, F-successful. The number of substitutions σ that is to be checked is exponential, but they can be enumerated in PSpace. Whether σ maps only to δ-inhabited transition can be tested in PSpace by Theorem 1. Computing $\delta^\sigma(pat(G))$ can be done in PTime by Lemma 2.

So far we have shown that INCL$_\Sigma$(*cPat*, NFA) is PSpace-complete. We next consider regular matching. This will permit us to show that INCL$_\Sigma$(*cPat*, DFA) is PSpace-hard too.

Proposition 2. MATCH$_\Sigma$(*cPat*, NFA) *is in* PSpace.

The proof is analogous to that of Proposition 1, except that it is now sufficient to guess some substitution to δ-inhabited transition.

Proposition 3. MATCH$_\Sigma$(*Pat*, DFA) *is* PSpace-*hard*.

The proof is by reduction from the non-emptiness problem of the intersection of a sequence of n DFAs $A_1, \ldots, A_n$ over Σ by matching the non-linear string pattern $y\# \ldots \#y$ of length n against a DFA A over $\Sigma \uplus \{\#\}$ recognizing $\mathcal{L}(A) = \mathcal{L}(A_1)\# \ldots \#\mathcal{L}(A_n)$.

Lemma 3. *The problem* coMATCH$_\Sigma$(*Pat*, DFA) *is* PSpace-*complete*.

In order to complete the proof of Theorem 2 it remains to relate regular inclusion and matching in the case of DFAs.

Lemma 4. $\text{COMATCH}_{\Sigma}(\mathcal{G}, \text{DFA}) =_p \text{INCL}_{\Sigma}(\mathcal{G}, \text{DFA})$ *for all $\mathcal{G}$ up to* PTIME *red.*

Proof. This follows from $Inst(pat(G)) \cap \mathcal{L}(A) = \emptyset \Leftrightarrow Inst(pat(G)) \subseteq \overline{\mathcal{L}(A)}$ and the fact that any DFA A can be complemented in PTIME to some DFA $\overline{A}$ such that $\overline{L(A)} = L(\overline{A})$. $\qquad\blacksquare$

The complexity of inclusion and matching decreases for linear string patterns, with or without compression. We indeed obtain the same complexity as known for streams (see Fig. 4), even though unknown factors and compression are permitted in addition.

Theorem 3 (Linear patterns). *Restricted to linear string patterns, regular compressed pattern inclusion and matching have the following complexities:*

1. $\text{INCL}_{\Sigma}(LinPat, \text{NFA})$ *and* $\text{INCL}_{\Sigma}(cLinPat, \text{NFA})$ *are* PSPACE-*complete if* $|\Sigma| \geq 2$ *while the problem* $\text{INCL}_{\Sigma}(cLinPat, \text{DFA})$ *can be solved in* PTIME.
2. $\text{MATCH}_{\Sigma}(cLinPat, \text{NFA})$ *is in* PTIME.

For any linear compressed string pattern G there exist a DFA of that recognizes $Inst(pat(G))$, but this DFA may be of exponential size in $|G|$ due to compression. Instead, the PTIME proof for $\text{INCL}_{\Sigma}(cLinPat, \text{DFA})$ relies on an evaluator of compressed string patterns in the transition monoid, that replaces free variables by the accessibility transition of the DFA, thereby testing a *reachability property*.

5   Defining Queries by Automata

We now recall the notion of queries on strings over some alphabet Σ with *variables* in some finite set $\mathcal{V}$ and relate them to languages of $\mathcal{V}$-annotated strings called $\mathcal{V}$-structures in [26]. We fix two disjoint finite sets Σ and $\mathcal{V}$.

Definition 2 (Query). *A query with variables in $\mathcal{V}$ on strings over Σ, or a Σ, $\mathcal{V}$-query for short, is a function $\mathbf{Q}$ that maps any string $w \in \Sigma^*$ to a set $\mathbf{Q}(w)$ of total assignments from $\mathcal{V}$ to $pos(w)$. A Boolean query is a Σ, $\emptyset$-query.*

Example 3. Let $\mathcal{V} = \{x, x'\}$. The query $\mathbf{Q}_1$ selects all pairs of letters (x, x') such that position x is labeled by a, position x' immediately follows x and is labeled by b. This query then satisfies $\mathbf{Q}_1(aa) = \emptyset$, $\mathbf{Q}_1(ab) = \{[x/1, x'/2]\}$, $\mathbf{Q}_1(abab) = \{[x/1, x'/2], [x/3, x'/4]\}$, etc.

We next show how a Σ, $\mathcal{V}$-query can be identified with a language of $\mathcal{V}$-annotated strings, i.e., of words over the alphabet $\Sigma^{\mathcal{V}} = \Sigma \times 2^{\mathcal{V}}$. A *(query variable) assignment* α to positions of a string $w \in \Sigma^*$ is a partial function from $\mathcal{V}$ to $pos(w)$. We will identify such variable assignments with words whose letters are sets of variables. For any partial function α from $\mathcal{V}$ to $pos(w)$ where $w \in \Sigma^n$, we define a corresponding word in $(2^{\mathcal{V}})^n$ by $word(\alpha) = \alpha^{-1}(1) \dots \alpha^{-1}(n)$. That is, $word(\alpha)[i]$ is the set of variables $x \in dom(\alpha)$ s.t. $\alpha(x) = i$. Furthermore, for any string $w \in \Sigma^*$ and variable assignment α into positions of w, we define the $\mathcal{V}$-annotated string $w * \alpha$ as a word over $\Sigma^{\mathcal{V}}$ by $w * \alpha = w * word(\alpha)$. In examples we will write a^V instead of letters $(a, V) \in \Sigma^{\mathcal{V}}$. For instance, $ab * [x/1, x'/2]$ is written as $a^{\{x\}} b^{\{x'\}}$.

Definition 3 ($\mathcal{V}$-structure [26]). *The set of $\mathcal{V}$-structures over Σ is the following set of $\mathcal{V}$-annotated strings, i.e., of words over $\Sigma^{\mathcal{V}}$:*

$$Struct^{\mathcal{V}} = \{w * \alpha \in (\Sigma^{\mathcal{V}})^* \mid w \in \Sigma^*, \ \alpha : \mathcal{V} \to pos(w)\}.$$

We note that all the assignments α in the definition of $\mathcal{V}$-structures are total functions. For instance for $\mathcal{V} = \{x, x'\}$ and $\Sigma = \{a, b\}$, the words $a^{\emptyset} b^{\{x', x\}}$ and $a^{\{x'\}} b^{\{x\}}$ are $\mathcal{V}$-structures while the words $a^{\emptyset} b^{\{x'\}}$ and $a^{\{x'\}} b^{\{x', x\}}$ are not. Essentially, $\mathcal{V}$-structures represent total variable assignments to the positions of a string without naming the positions.

Definition 4 (Language of $\mathcal{V}$-structure of a query). *For any Σ, $\mathcal{V}$-query $\boldsymbol{Q}$, the* language of $\mathcal{V}$-structures *of $\boldsymbol{Q}$ is $\mathcal{L}(\boldsymbol{Q}) = \{w * \alpha \mid w \in \Sigma^*, \ \alpha \in \boldsymbol{Q}(w)\}.$*

We will be interested in queries whose languages are definable by NFAs.

Definition 5 (Query automata). *An Σ, $\mathcal{V}$-query automaton is an NFA A such that $\mathcal{L}(A)$ is a language of $\mathcal{V}$-structures over Σ. The unique Σ, $\mathcal{V}$-query such that $\mathcal{L}(\boldsymbol{Q}) = \mathcal{L}(A)$ is called the query defined by A and is denoted by $\mathcal{Q}(A) = \boldsymbol{Q}$.*

6   Certain Query Answers and Non-answers

We next formalize the notions of certain query answers and non-answers on string patterns. For streams, these definitions coincide with the notions of earliest query answers from [12] and fast-failure from [4], respectively.

A Σ-*assignment* for $\mathcal{V}$ on $p \in Pat_{\Sigma}$ is a partial function α from $\mathcal{V}$ to $pos_{\Sigma}(p)$, the Σ-positions of the pattern p. For any Σ-assignment α on p, the word $p * \alpha$ is a string pattern over $\Sigma^{\mathcal{V}}$, still with string variables in $\mathcal{Y}$. Therefore, the set $Inst(p * \alpha)$ is a well-defined set of words over $\Sigma^{\mathcal{V}}$. Note, however, that some of these $\mathcal{V}$-annotated strings may not be $\mathcal{V}$-structures. For instance, if $x \in \mathcal{V}$, $a \in \Sigma$, and $Y \in \mathcal{Y}$, then $a^{\{x\}} a^{\{x\}} \in Inst(Y * [])$ is not a $\mathcal{V}$-structure since x occurs twice (where $[]$ is the empty Σ-assignment).

Definition 6 (Certain query answers and non-answers). *Let $\boldsymbol{Q}$ be a Σ, $\mathcal{V}$-query, and let $p \in Pat_{\Sigma}$ be a string pattern. A Σ-assignment α for $\mathcal{V}$ on p is:*

- *a* certain answer *for query $\boldsymbol{Q}$ on p if α is total and $Inst(p * \alpha) \cap Struct^{\mathcal{V}} \subseteq \mathcal{L}(\boldsymbol{Q})$,*
- *and a* certain non-answer *for query $\boldsymbol{Q}$ on p if $Inst(p * \alpha) \cap \mathcal{L}(\boldsymbol{Q}) = \emptyset$.*

Given an instance $w \in Inst(p)$, each Σ-assignment α on p defines a set of total Σ-assignments on w, where all variables not in $dom(\alpha)$ must be mapped to some Σ-positions "created" by the instantiation. More formally:

$$C_{p,w}(\alpha) = \{\alpha' \mid \alpha' \text{ is a total } \Sigma\text{-assignment on } w, \ w * \alpha' \in Inst(p * \alpha)\}.$$

The offsets of positions of query variables due to the instantiation of pattern variables raise two issues that we illustrate in the following example: 1. even a total Σ-assignment α of p might have several completions for the same string w, and 2. it might be the case that $\alpha \notin C_{p,w}(\alpha)$.

Example 4. Consider the string pattern $p = ay_1ay_2$, the string $w = aaba$ in $Inst(p)$, $\mathcal{V} = \{x\}$, and the total Σ-assignment $\alpha = [x/3]$ on p. In order to make p match w, the second a-letter of p matches either the second or the fourth position in w. Therefore, there are exactly two substitutions that make p match w, which are $\sigma_1 = [y_1/\epsilon, y_2/ba]$ and $\sigma_2 = [y_1/ab, y_2/\epsilon]$. Now, $\hat{\sigma_1}(p * \alpha) = \hat{\sigma_1}(ay_1a^{\{x\}}y_2) = aa^{\{x\}}ba = w * [x/2]$, thus $[x/2] \in C_{p,w}(\alpha)$. Also, $\hat{\sigma_2}(p * \alpha) = \hat{\sigma_1}(ay_1a^{\{x\}}y_2) = aaba^{\{x\}} = w * [x/4]$, thus $[x/4] \in C_{p,w}(\alpha)$. Given that there is no further way to match p with w, there is no further completion. That is, $C_{p,w}(\alpha) = \{[x/2], [x/4]\}$, and $\alpha \notin C_{p,w}(\alpha)$.

The next proposition relates certain query answers (resp. non-answers) on a pattern p to query answers (resp. non-answers) on its instances.

Proposition 4. *Let α be a Σ-assignment on string pattern p and $\boldsymbol{Q}$ be a Σ, $\mathcal{V}$-query. It then holds for all instances $w \in Inst(p)$:*

- *If α is a certain answer for query $\boldsymbol{Q}$ on p then $C_{p,w}(\alpha) \subseteq \boldsymbol{Q}(w)$.*
- *If α is a certain non-answer for query $\boldsymbol{Q}$ on p then $C_{p,w}(\alpha) \cap \boldsymbol{Q}(w) = \emptyset$.*

7   Certain Query Answering and Non-answering

We introduce the problems of certain query answering and non-answering for classes of compressed string patterns $\mathcal{G}$ and of query NFAs $\mathcal{A}$:

Certain query answering. $\mathrm{CERT}_{\Sigma,\mathcal{V}}^{ans}(\mathcal{G}, \mathcal{A})$. *Input:* a compressed string pattern $G \in \mathcal{G}_\Sigma$, a Σ-assignment α for $\mathcal{V}$ on $pat(G)$, and a query NFA $A \in \mathcal{A}_{\Sigma\mathcal{V}}$.
  Output: whether α is a certain answer for query $\mathcal{Q}(A)$ on $pat(G)$.
Certain query non-answering. $\mathrm{CERT}_{\Sigma,\mathcal{V}}^{\neg ans}(\mathcal{G}, \mathcal{A})$. *Input:* as above
  Output: whether α is a certain non-answer for query $\mathcal{Q}(A)$ on $pat(G)$.

Note that α is an assignment to positions of $pat(G)$ and not to positions of G. This is necessary because due to compression, a position of G might correspond to several positions of the underlying word which need to be distinguished. Indeed, considering G to be the compressed string pattern in the introduction, the a-letter in the rule for X corresponds to positions 5 and 13 in the decompressed pattern $aYcZabbYaYcZa$, and position 5 is a certain answer for query $\boldsymbol{Q}_1$, while position 13 is a certain non-answer.

The following straightforward lemma relates certain query answering to regular inclusion and certain query non-answering to regular matching.

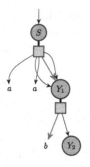
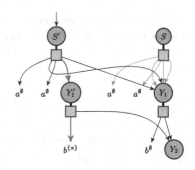

Fig. 7. G with red thick path to the shared position 5. (Color figure online)

Fig. 8. G' on $\Sigma^{\mathcal{V}}$ with $pat(G') = pat(G) * [x/5]$.

Lemma 5 (Boolean Queries). *For any classes $\mathcal{G}$ and $\mathcal{A}$, $\mathrm{INCL}_{\Sigma}(\mathcal{G}, \mathcal{A}) =_p$ $\mathrm{CERT}_{\Sigma,\emptyset}^{ans}(\mathcal{G}, \mathcal{A})$ and $\mathrm{COMATCH}_{\Sigma}(\mathcal{G}, \mathcal{A}) =_p \mathrm{CERT}_{\Sigma,\emptyset}^{\neg ans}(\mathcal{G}, \mathcal{A})$.*

Since PSPACE-complete problems are closed by complement, Lemma 5 implies together with Theorem 2 that $\mathrm{CERT}_{\Sigma,\emptyset}^{ans}(Pat, \mathrm{DFA})$ and $\mathrm{CERT}_{\Sigma,\emptyset}^{\neg ans}$ (Pat, DFA) are PSPACE-complete, even though these problems are restricted to DFAs, without compression, and for Boolean queries. Therefore, all certainty problems $\mathrm{CERT}_{\Sigma,\emptyset}^{\mathcal{B}}(\mathcal{G}, \mathcal{A})$ where $\mathcal{B} \in \{\neg ans, ans\}$, $\mathcal{G} \in \{Pat, cPat\}$, and $\mathcal{A} \in \{\mathrm{DFA}, \mathrm{NFA}\}$ are PSPACE-hard. In the sequel we show that all these problems can be solved in PSPACE for arbitrary finite sets Σ and $\mathcal{V}$. Basically, these results will be corollaries of Theorem 2, Lemma 5 on Boolean queries, and the following partial decompression lemma. This result is new to the best of our knowledge, even though its proof relies on similar techniques as used for instance in [5] for computing in PTIME the letter at the n-th position of $pat(G)$ for a singleton grammar G.

Lemma 6 (Partial Decompression). *For any $G \in cPat_{\Sigma}$ and any Σ-assignment α for $\mathcal{V}$ on $pat(G)$, we can compute in PTIME some $G' \in cPat_{\Sigma^{\mathcal{V}}}$ such that $pat(G) * \alpha = pat(G')$. In particular, if $pat(G)$ was linear then $pat(G')$ is linear.*

For illustration, let $\mathcal{V} = \{x\}$, $\alpha = [x/5]$. Figure 8 shows a compressed string pattern G' obtained by partially decompressing the compressed string pattern G in Fig. 7 at the position 5 of its pattern, such that $pat(G) * \alpha = pat(G')$.

Proposition 5. *For all $\mathcal{B}$ in $\{ans, \neg ans\}$, all $\mathcal{G}$ in $\{Pat, cPat, LinPat, cLinPat\}$, and all $\mathcal{A}$ in $\{\mathrm{DFA}, \mathrm{NFA}\}$, there is a PTIME reduction from $\mathrm{CERT}_{\Sigma,\mathcal{V}}^{\mathcal{B}}(\mathcal{G}, \mathcal{A})$ to $\mathrm{CERT}_{\Sigma^{\mathcal{V}},\emptyset}^{\mathcal{B}}(\mathcal{G}, \mathcal{A})$.*

Corollary 1 (Non-linear patterns). *For all $\mathcal{B} \in \{ans, \neg ans\}$, $\mathcal{G} \in \{Pat, cPat\}$ and $\mathcal{A} \in \{\mathrm{DFA}, \mathrm{NFA}\}$, the problem $\mathrm{CERT}_{\Sigma,\mathcal{V}}^{\mathcal{B}}(\mathcal{G}, \mathcal{A})$ is PSPACE-complete.*

Corollary 2 (Linear patterns). *For all* $\mathcal{A} \in \{\text{DFA}, \text{NFA}\}$, *the certainty problems* $\text{CERT}_{\Sigma,\mathcal{V}}^{\neg ans}(cLinPat, \mathcal{A})$ *(1) and* $\text{CERT}_{\Sigma,\mathcal{V}}^{ans}(cLinPat, \text{DFA})$ *(2) can be solved in* PTIME. *The problem* $\text{CERT}_{\Sigma,\mathcal{V}}^{ans}(\mathcal{G}, \text{NFA})$ *is* PSPACE-*complete for all* $\mathcal{G} \in \{LinPat, cLinPat\}$ *(3).*

8 Conclusion

There exist highly efficient streaming algorithms for answering queries defined by NFAs [10] on complex event streams with low latency, but not with lowest latency, since they approximate the sets of certain query answers at any event. The positive results presented here yield good hope that similar algorithms could be developed for hyperstreams when approximated by compressed linear string patterns. As shown by the authors in a follow-up paper, the linearity restriction is not sufficient. But with a further restriction it is possible to approximate certain query answers efficiently and with decent precision [8]. However, this still requires more research. First, one needs to understand how such algorithms may deal with unknown factors incrementally, without requiring cubic time per step such as previous algorithms on incremental evaluation of queries defined by NFAs [6]. Second, one has to understand how to deal with nested word automata rather than NFAs for dealing with regular path queries on complex event streams. Another point is to develop streaming algorithms for hyperstreams with data values from an infinite signature. Finally, the feasibility of hyperstreaming algorithms needs to be proven in practice.

Acknowledgments. We are thankful to C. Paperman, who saw the PSPACE-hardness of regular string pattern matching in a discussion on the topic. We thank S. Salvati and S. Tison for discussions on regular string pattern matching. It is a pleasure to thank all the anonymous reviewers for their extraordinary helpful feedback.

References

1. Angles, R., Arenas, M., Barceló, P., Hogan, A., Reutter, J.L., Vrgoc, D.: Foundations of modern graph query languages. CoRR, abs/1610.06264 (2016)
2. Angluin, D.: Finding patterns common to a set of strings. J. Comput. Syst. Sci. **21**, 46–62 (1980)
3. Babai, L., Szemeredi, E.: On the complexity of matrix group problems i. In: Proceedings of the 25th Annual Symposium on Foundations of Computer Science, SFCS 1984, pp. 229–240. IEEE Computer Society, Washington, DC (1984)
4. Benedikt, M., Jeffrey, A., Ley-Wild, R.: Stream firewalling of XML constraints. In: ACM SIGMOD International Conference on Management of Data, pp. 487–498. ACM-Press (2008)
5. Bille, P., Landau, G.M., Raman, R., Sadakane, K., Satti, S.R., Weimann, O.: Random access to grammar-compressed strings and trees. SIAM J. Comput. **44**(3), 513–539 (2015)
6. Björklund, H., Gelade, W., Martens, W.: Incremental XPath evaluation. ACM Trans. Database Syst. **35**(4), 29 (2010)

7. Blondin, M., Krebs, A., McKenzie, P.: The complexity of intersecting finite automata having few final states. Comput. Complex. **25**(4), 775–814 (2016)
8. Boneva, I., Niehren, J., Sakho, M.: Approximating certain query answering on hyperstreams. Technical report, June 2018
9. David, C., Libkin, L., Murlak, F.: Certain answers for XML queries. In: Proceedings of the Twenty-Ninth ACM SIGMOD-SIGACT-SIGART Symposium on Principles of Database Systems, PODS 2010, Indianapolis, Indiana, USA, 6–11 June 2010, pp. 191–202. ACM (2010)
10. Debarbieux, D., Gauwin, O., Niehren, J., Sebastian, T., Zergaoui, M.: Early nested word automata for XPath query answering on XML streams. Theor. Comput. Sci. **578**, 100–125 (2015)
11. Gascón, A., Godoy, G., Schmidt-Schauß, M.: Context matching for compressed terms. In: Proceedings of the Twenty-Third Annual IEEE Symposium on Logic in Computer Science, LICS 2008, Pittsburgh, PA, USA, 24–27 June 2008, pp. 93–102. IEEE Computer Society (2008)
12. Gauwin, O., Niehren, J.: Streamable fragments of forward XPath. In: Bouchou-Markhoff, B., Caron, P., Champarnaud, J.-M., Maurel, D. (eds.) CIAA 2011. LNCS, vol. 6807, pp. 3–15. Springer, Heidelberg (2011). https://doi.org/10.1007/978-3-642-22256-6_2
13. Gauwin, O., Niehren, J., Tison, S.: Earliest query answering for deterministic nested word automata. In: Kutyłowski, M., Charatonik, W., Gębala, M. (eds.) FCT 2009. LNCS, vol. 5699, pp. 121–132. Springer, Heidelberg (2009). https://doi.org/10.1007/978-3-642-03409-1_12
14. Gauwin, O., Niehren, J., Tison, S.: Queries on XML streams with bounded delay and concurrency. Inf. Comput. **209**, 409–442 (2011)
15. Green, T.J., Gupta, A., Miklau, G., Onizuka, M., Suciu, D.: Processing XML streams with deterministic automata and stream indexes. ACM Trans. Database Syst. **29**(4), 752–788 (2004)
16. Kay, M.: A streaming XSLT processor. In: Balisage: The Markup Conference 2010. Balisage Series on Markup Technologies, vol. 5 (2010)
17. Kozen, D.: Lower bounds for natural proof systems. In: 18th Annual Symposium on Foundations of Computer Science, Providence, Rhode Island, USA, 31 October - 1 November 1977, pp. 254–266. IEEE Computer Society (1977)
18. Kumar, V., Madhusudan, P., Viswanathan, M.: Visibly pushdown automata for streaming XML. In: 16th International Conference on World Wide Web, pp. 1053–1062. ACM-Press (2007)
19. Kupferman, O., Vardi, M.Y.: Model checking of safety properties. Form. Methods Syst. Des. **19**(3), 291–314 (2001)
20. Labath, P., Niehren, J.: A functional language for hyperstreaming XSLT. Technical report, INRIA Lille (2013)
21. Maneth, S., Ordóñez, A., Seidl, H.: Transforming XML streams with references. In: Iliopoulos, C., Puglisi, S., Yilmaz, E. (eds.) SPIRE 2015. LNCS, vol. 9309, pp. 33–45. Springer, Cham (2015). https://doi.org/10.1007/978-3-319-23826-5_4
22. Mozafari, B., Zeng, K., Zaniolo, C.: High-performance complex event processing over XML streams. In: Candan, K.S., et al. (eds.) SIGMOD Conference, pp. 253–264. ACM (2012)
23. Olteanu, D.: SPEX: streamed and progressive evaluation of XPath. IEEE Trans. Know. Data Eng. **19**(7), 934–949 (2007)
24. Plandowski, W.: The complexity of the morphism equivalence problem for context-free languages. Ph.D. thesis. Department of Informatics, Mathematics, and Mechanics, Warsaw University (1995)

25. Schmidt, M., Scherzinger, S., Koch, C.: Combined static and dynamic analysis for effective buffer minimization in streaming XQuery evaluation. In: 23rd IEEE International Conference on Data Engineering, pp. 236–245 (2007)
26. Straubing, H.: Finite Automata, Formal Logic and Circuit Complexity. Progress in Computer Science and Applied Series. Birkhäuser, Basel (1994)

Büchi VASS Recognise Σ_1^1-complete ω-languages

Michał Skrzypczak[✉][ID]

University of Warsaw, Warsaw, Poland
mskrzypczak@mimuw.edu.pl

Abstract. This paper exhibits an example of a Σ_1^1-complete ω-language that can be recognised by a Büchi automaton with one partially blind counter (or equivalently a Büchi VASS with only one place). It follows as a corollary that there is no equivalent model of deterministic automata, even if we allow much richer data structures than just counters. The same holds for weaker forms of determinism, like for unambiguous or countably-unambiguous machines. This shows that even in the one counter case, non-determinism of Büchi VASS is inherent.

Keywords: Petri nets · Infinite words · Non-determinism

In this work we study the strength of non-determinism in the context of partially blind multi-counter Büchi automata. This is a model of finite automata over infinite words with the Büchi acceptance condition (also known as "repeated reachability condition"). Additionally, each such automaton is equipped with a finite set of counters taking non-negative integer values. The automaton can freely increment and decrement the values of the counters, however it cannot test these values (i.e. no zero nor equality test). The only way in which the values of the counters influence the behaviour of the automaton is that they must stay non-negative during a run. The studied class of automata is strongly connected with other models based on Petri nets: a partially blind multi-counter Büchi automaton can be seen as a Büchi Vector Addition System with States (i.e. Büchi VASS) and vice versa.

Similarly as in the case of Petri nets, the considered model is naturally equipped with non-determinism. The main result of [10] implies that Büchi VASS are able to recognise ω-languages that cannot be recognised by the deterministic variant of the machines. This was achieved by topological methods: the paper provides an example of a Büchi VASS recognising an ω-language complete for the third level of the hierarchy of Borel sets (Σ_3^0-complete); while deterministic Büchi VASS can only recognise ω-languages in the second level of the hierarchy (in Π_2^0).

This work has been supported by Poland's National Science Centre (NCN) grant no. 2016/21/D/ST6/00491.

I. Potapov and P.-A. Reynier (Eds.): RP 2018, LNCS 11123, pp. 133–145, 2018.
https://doi.org/10.1007/978-3-030-00250-3_10

While the result of [10] separates non-deterministic Büchi VASS from the deterministic ones, it does not settle the question of the upper bounds on the topological complexity for these machines. Moreover, the lower bound of Σ_3^0 does not rule out the possibility of having a model of automata with a limited form of non-determinism that still captures the expressive power of non-deterministic Büchi VASS. To counterbalance the lack of full non-determinism, one could consider adding new counter operations (like min and max, see e.g. [1,3]); extending the acceptance condition to a topologically harder one, like Rabin, lim inf-parity, or something like the ωBS condition from [2]; or adding a richer data structure, e.g. a stack. Also, instead of a fully deterministic model, one could hope for an intermediate form of non-determinism, as in the case of unambiguous machines [5]; or when the non-deterministic choices appear only finitely many times in the accepting runs. The latter assumption implies that there are at most countably many accepting runs over a fixed ω-word, we will call such a machine *countably-unambigous*. This last restriction finds justification in the actual example provided in [10], where the whole non-deterministic choice of the machine reduces to choosing a single natural number at the beginning of a run.

In general, topological complexity suits well to make a distinction between determinism and non-determinism. Firstly, in the case of all standard models of machines, the relation $\mathrm{run}(\alpha, \rho)$ of "being a run" is a closed[1] relation between ω-words $\alpha \in A^\omega$ and sequences of configurations $\rho \in C^\omega$. Moreover, for all the standard acceptance conditions mentioned above, the property of being an accepting run $\mathrm{acc}(\rho)$ is Borel. This implies that all deterministic devices, which can be seen as transducers of an input ω-word into a sequence of configurations, recognise only Borel sets. The situation is different in the case of non-deterministic devices, where the language of such a machine can be written as a projection of a Borel set:

$$\{\alpha \in A^\omega \mid \exists \rho \in C^\omega.\ \mathrm{run}(\alpha, \rho) \wedge \mathrm{acc}(\rho)\} = \pi_{A^\omega}\left(\{(\alpha, \rho) \mid \mathrm{run}(\alpha, \rho) \wedge \mathrm{acc}(\rho)\}\right). \quad (1)$$

It is known that in general, projections of Borel sets might not be Borel—they form a wider class of *analytic* sets (denoted Σ_1^1). Thus, Σ_1^1 is the upper bound for the topological complexity of general non-deterministic devices. The above formula, together with a theorem by Lusin and Novikov [12, Theorem 18.10], imply that countably-unambiguous machines recognise only Borel ω-languages. This means that in terms of topological complexity they are closer to deterministic than to non-deterministic ones.

The above topological results say that the distinction between weak vs. full forms of non-determinism can be topologically understood as the difference between Borel and analytic sets. The purpose of the present paper is to use this correspondence by showing the following theorem.

Theorem 1. *There exists an ω-language that is recognised by a Büchi VASS with one counter (i.e. with one place) that recognises a Σ_1^1-complete ω-language.*

---

[1] Equivalently: a relation given by a *safety* condition.

As noted above, all ω-languages recognised by non-deterministic Büchi VASS are in Σ_1^1. Thus, the above result solves the question of the upper bounds for the topological complexity of these machines. Moreover, the theorem translates to the automata theoretic realms as the following corollary.

Corollary 1. *No model of deterministic, unambiguous, nor even countably-unambiguous automata with countably many configurations and a Borel acceptance condition can capture the class of ω-languages recognisable by Büchi VASS with one counter.*

The crucial difficulty in proving Theorem 1 is the fact, that Büchi VASS are partially blind: they cannot test their counters for exact values. As a consequence, there is a natural simulation order on the configurations of a Büchi VASS: a configuration $(q, \vec{a})$ *simulates* $(q, \vec{c})$ if they have the same state and the counter values A and B satisfy coordinate-wise $\vec{a} \geq \vec{c}$. In such a case, the language recognised from $(q, \vec{a})$ contains the language recognised by $(q, \vec{c})$; because each accepting run from $(q\vec{c})$ can be lifted to an accepting run from $(q, \vec{a})$ just by increasing the counter values. In particular, when there is exactly one counter, the maximal size of an anti-chain of the simulation order is bounded by the number of states; what limits the possible structure of the so-called *residual* ω-languages of the device.

Although the construction of the paper is expressed in terms of topological complexity, the actual core of the proof is a combinatorial idea allowing to simulate a Σ_1^1-hard behaviour (i.e. one that involves full non-determinism) by an efficient way of storing information in the value of a unique partially blind counter of the automaton. The idea is not very complex, and the overall construction should be considered as rather direct.

To simplify the presentation of the proof it is performed in three steps. In Sect. 2 we provide an easy example of a Σ_1^1-complete ω-language recognised by a Büchi VASS with two counters. Then, in Sect. 3 we characterise a specific Σ_1^1-complete set (namely $\mathrm{IF}_{\mathrm{inf}}$). This set is in a certain sense monotone, which is used to reflect the simulation order on configurations of our automata. In Sect. 4 we reduce the set $\mathrm{IF}_{\mathrm{inf}}$ to an ω-language recognised by a Büchi VASS with only one counter, which concludes the proof of Theorem 1. Section 5 is devoted to Corollary 1. Finally, Sect. 6 gives some concluding remarks.

Acceptance Condition. The results of the paper speak about VASS with the Büchi acceptance condition. Since non-deterministic Büchi automata recognise all ω-regular languages, these machines can simulate all other ω-regular acceptance conditions. Thus, the Büchi condition seems to be one of the canonical ones (with most of them actually equivalent). On the other hand, the situation is different for certain weaker acceptance conditions: the safety, reachability, and co-Büchi conditions can be written as countable unions of closed sets (i.e. Σ_2^0). A known topological fact says that a projection of a Σ_2^0 set contained in a compact[2] topological space is also $\mathbf{\Sigma}_2^0$. Therefore, none of these weaker conditions allows a non-deterministic VASS to recognise a non-Borel ω-language. A reasonable task (although out of the scope of the present paper) is to design deterministic or almost deterministic models for VASS with these weaker acceptance conditions.

Related Work. There is a number of papers studying the topological complexity of sets recognisable by various models of machines [4,6–8,13]. In certain cases, the topological lower bounds were used to separate models of machines [1,10]. Also, high topological complexity of some classes of languages can influence their decidability [11].

The question of upper bounds on the topological complexity for Büchi VASS was left as an open problem in [10]. After publication of that article, the authors independently managed to solve this problem. In [9], Finkel has found a family of Büchi VASS with four counters that recognise ω-languages at all Wadge degrees of non-deterministic Turing machines. This result implies that there are Büchi VASS with 4 counters recognising $\mathbf{\Sigma}_1^1$-complete ω-languages. Moreover, it shows that many intermediate classes of topological complexity are also inhabited by such ω-languages. However, it is not clear whether the number of counters in that construction can be reduced. This paper provides a construction of a single $\mathbf{\Sigma}_1^1$-complete ω-language recognised using only one counter. Thus, the two results are mathematically incomparable.

1   Preliminary Notions

We use $\omega = \{0, 1, \ldots\}$ to denote the set of natural numbers. If A is a non-empty set then A^* and A^ω are respectively sets of finite and infinite sequences of elements of A. The elements of A^* are called *words* and the elements of A^ω are called ω-words. An ω-*language* is a set of ω-words. If $v \in A^*$ then by $|v| \in \omega$ we denote the length of v (i.e. the number of symbols in v). By $v \cdot x$ we denote the concatenation of the two sequences, with $|v \cdot x| = |v| + |x|$. If the context is clear, we skip the concatenation symbol $\cdot$. If $n \leq |v|$ then by $v{\restriction}_n \in A^n$ we denote the restriction of the sequence to its first n symbols.

Büchi VASS. A Büchi VASS (or shortly VASS, as we consider only the Büchi acceptance condition) is a tuple $\mathcal{A} = \langle A, Q, q_\mathrm{I}, F, C, \delta \rangle$, where:

- A is a finite *input alphabet*,
- Q is a finite set of *states*,
- $q_\mathrm{I} \in Q$ is an *initial state*,
- $F \subseteq Q$ is a set of *accepting states*,
- C is a finite set of *counters*,
- δ is a finite *transition relation*, its elements are *transitions* (q, a, τ, q') where $q, q' \in Q$, $a \in A$, and $\tau : C \to \mathbb{Z}$.

---

[2] The space of runs is compact because the automata do not admit ϵ-transitions and therefore the possible counter values are bounded at each fixed place of the input ω-word.

Without loss of generality we assume that the set of counters C has the form $C = \{1, 2, \ldots, k\}$ for some k (in this work 1 or 2). We visually represent a transition (q, a, τ, q') by $q \xrightarrow{a:\big(\tau(1),\tau(2),\ldots,\tau(k)\big)} q'$. We say that such a transition is *over* the letter a. If $A' \subseteq A$ then $q \xrightarrow{A':\big(\tau(1),\tau(2),\ldots,\tau(k)\big)} q'$ means that for each $a \in A'$ there is a respective transition. Similarly, $q \xrightarrow{a} q'$ and $q \xrightarrow{A'} q'$ denote the respective transitions that do not modify the counter values (i.e. τ is constant 0).

A *configuration* of a VASS $\mathcal{A}$ is a tuple $(q, c_1, c_2, \ldots, c_k)$ where $q \in Q$, $c_1, \ldots, c_k \in \omega$, and $\{1, \ldots, k\} = C$. The *initial configuration* is $(q_I, 0, \ldots, 0)$. We say that a transition $q \xrightarrow{a:\big(\tau(1),\ldots,\tau(k)\big)} q'$ *goes* from a configuration $(q, c_1, \ldots, c_k)$ to a configuration $\big(q', c_1 + \tau(1), \ldots, c_k + \tau(k)\big)$ (note that by the definition it requires all the numbers $c_i + \tau(i)$ to be non-negative).

Let $\alpha \in A^\omega$ be an ω-word over the input alphabet. A *run* of a VASS $\mathcal{A}$ over α is an infinite sequence ρ of configurations, such that $\rho(0)$ is the initial configuration and for every $i \in \omega$ there is a transition of $\mathcal{A}$ over the letter $\alpha(i)$ that goes from the configuration $\rho(i)$ to the configuration $\rho(i+1)$. A run ρ is *accepting* if for infinitely many i the configuration $\rho(i) = (q_i, \ldots)$ satisfies $q_i \in F$ (i.e. it visits infinitely many times an accepting state). A VASS $\mathcal{A}$ *accepts* an ω-word α if there exists an accepting run of $\mathcal{A}$ over α. The *language* of $\mathcal{A}$ (denoted $L(\mathcal{A})$) is the set of ω-words accepted by $\mathcal{A}$.

Topology. We will use the standard notions of topology on Polish spaces [12]. The space A^ω of all ω-words over a finite alphabet A can be naturally endowed with a topology where open sets are those obtained as unions of *basic open* sets of the form $N_u \stackrel{\text{def}}{=} \{u \cdot \alpha \mid \alpha \in A^\omega\}$. A set whose complement is open is called *closed*. Closed subsets C of A^ω can be equivalently characterised as those satisfying the following *safety* property:

$$\forall \alpha \in A^\omega. \ \big(\forall n \in \omega . \exists \beta \in A^\omega . \alpha\!\restriction_n \cdot \beta \in C\big) \implies \alpha \in C. \tag{2}$$

The family of Borel sets in a topological space X is the smallest σ-algebra that contains all the open sets in X. By Σ_1^1 we denote the family of analytic sets, i.e. projections of Borel sets. A function $f \colon X \to Y$ between two topological spaces is *continuous* if the pre-image $f^{-1}(U) \subseteq X$ is open for every open[3] set $U \subseteq Y$. If $A \subseteq X$ and $B \subseteq Y$ are two subsets of topological spaces then we call $f \colon X \to Y$ a *reduction* of A to B if $f^{-1}(B) = A$. If Γ is a class of sets and $G \subseteq X$ is a subset of a topological space X, we say that G is Γ-*hard* if for every set $A \in \Gamma$ there exists a continuous reduction of A to G. If additionally $G \in \Gamma$ then we say that G is Γ-*complete*. Since continuous reductions can be composed, we obtain the following fact.

Fact 2. *If G is Γ-hard and G continuously reduces to G' then also G' is Γ-hard.*

Orders. Consider a set X and a relation $o \subseteq X \times X$ on X. We say that o is a *linear order* if it is reflexive, transitive, and anti-symmetric. We interpret a pair

---

[3] Since $f^{-1}\big(\bigcup \mathcal{F}\big) = \bigcup f^{-1}(\mathcal{F})$, it is enough to consider basic open sets U.

$(x, x') \in o$ as representing the fact that x is o-smaller-or-equal than x'. A linear order o is *ill-founded* if there exists an infinite sequence $x_0, x_1, \ldots$ of pairwise distinct elements of X such that for all n we have $(x_{n+1}, x_n) \in o$. Such a sequence indicates an infinite o-descending chain. An order that is not ill-founded is called *well-founded*.

Binary Trees. The binary tree is the set of all sequences of *directions* $\mathcal{T} \stackrel{\text{def}}{=} \{L, R\}^*$ where the *directions* L, R are two fixed distinct symbols. For technical reasons we sometimes consider a third direction M (it does not occur in the binary tree).

A set $X \subseteq \mathcal{T}$ can be naturally identified with its characteristic function $X \in \{0, 1\}^{\left(\{L,R\}^*\right)}$. Thus, the family of all subsets of the binary tree, with the natural product topology, is homeomorphic with the Cantor set $\{0, 1\}^\omega$.

The elements $v, x \in \mathcal{T}$ are called *nodes*. Nodes are naturally ordered by the following three orders:

- the prefix order: $v \preceq x$ if x can be obtained by concatenating something at the end of v,
- the lexicographic order: $v \leq_{\text{lex}} x$ if v is lexicographically smaller than x (we assume that $L <_{\text{lex}} M <_{\text{lex}} R$),
- the infix order: $v \leq_{\text{inf}} x$ if $v M^\omega$ (i.e. the ω-word obtained by appending infinitely many symbols M after v) is lexicographically less or equal than $x M^\omega$.

Notice that, for every fixed n, when restricted to $\{L, R\}^n$, the lexicographic and infix orders coincide. However, $L <_{\text{inf}} \epsilon <_{\text{inf}} R$ but ϵ is the minimal element of $\leq_{\text{lex}}$. Both the lexicographic and infix orders are linear.

Since the infix order is countable, dense, and has no minimal nor maximal elements, we obtain the following fact.

Fact 3. $(\mathcal{T}, \leq_{\text{inf}})$ *is isomorphic to the order of rational numbers* $(\mathbb{Q}, \leq)$.

Hardness. In the following part of the paper we will use the following two sets:

$$\text{IF}_{\text{pre}} \stackrel{\text{def}}{=} \{X \subseteq \mathcal{T} \mid X \text{ contains an infinite } \preceq\text{-ascending chain}\},$$

$$\text{IF}_{\text{inf}} \stackrel{\text{def}}{=} \{X \subseteq \mathcal{T} \mid X \text{ contains an infinite } \leq_{\text{inf}}\text{-descending chain}\}.$$

The following lemma is a standard topological observation.

Lemma 1. *The sets* IF_{pre} *and* IF_{inf} *are* Σ_1^1*-complete.*

Proof. Both sets belong to Σ_1^1 just by the form of the definition. IF_{pre} is Σ_1^1-hard by an easy reduction from the set of ill-founded ω-branching trees, the proof is similar to [12, Exercise 27.3].

IF_{inf} is Σ_1^1-hard by a reduction from the set of ill-founded linear orders on ω (seen as elements of $\{0, 1\}^{\omega \times \omega}$). Let us prove this fact more formally. Consider an element $o \in \{0, 1\}^{\omega \times \omega}$ that is a linear order on ω. The latter set is Σ_1^1-complete by a theorem by Lusin and Sierpiński [12, Theorem 27.12]. We will inductively define $X_o \subseteq \mathcal{T}$ in such a way to ensure that $o \mapsto X_o$ is a continuous mapping and o is ill-founded if and only if $X_o \in \mathrm{IF}_{\inf}$.

Let us proceed inductively, defining a sequence of nodes $(x_n)_{n \in \omega} \subseteq \mathcal{T}$. Our invariant says that $|x_k| = k$ and the map $k \mapsto x_k$ is an isomorphism of the orders $(\{0, 1, \ldots, n\}, o)$ and $(\{x_0, x_1, \ldots, x_n\}, \leq_{\inf})$. We start with $x_0 = \epsilon$ (i.e. the root of $\mathcal{T}$). Assume that $x_0, \ldots x_n$ are defined and satisfy the invariants. By the definition of $\leq_{\inf}$, there exists a node $x \in \{\mathtt{L}, \mathtt{R}\}^{n+1}$ such that for $k = 0, 1, \ldots, n$ we have $x \leq_{\inf} x_k$ if and only if $(n+1, k) \in o$. Let x_{n+1} be such a node.

The above induction defines an infinite sequence of nodes $x_0, x_1, \ldots$ Let $X_o \stackrel{\text{def}}{=} \{x_n \mid n \in \omega\} \subseteq \mathcal{T}$. By the definition of X_o, the mapping $o \mapsto X_o$ is continuous—the fact whether a node $x \in \mathcal{T}$ belongs to X_o depends only on $o \cap \{0, 1, \ldots, |x|\}^2$. Using our invariant, we know that the map $k \mapsto x_k$ is an isomorphism of the orders (ω, o) and $(X_o, \leq_{\inf})$. Thus, o is ill-founded if and only if $X_o \in \mathrm{IF}_{\inf}$. $\square$

2   Hardness for Two Counters

In this section we provide a simple example of an ω-language that is Σ_1^1-complete and can be recognised by a VASS $\mathcal{A}_2$ with two counters. This example should be seen as a preliminary step towards the one counter example given in Sect. 4.

The VASS $\mathcal{A}_2$ is depicted in Fig. 1. Let $A_0 \stackrel{\text{def}}{=} \{<, d_1, d_2, |, i_1, i_2, +, -, >\}$ and let the alphabet $A \stackrel{\text{def}}{=} A_0 \cup \{\sharp\}$. The initial state is q_0, the single accepting state is q_a. The only non-determinism occurs in q_0 when reading $<$—the VASS can stay in q_0 or move to q_1. Only the states q_1 and q_2 modify the counter values.

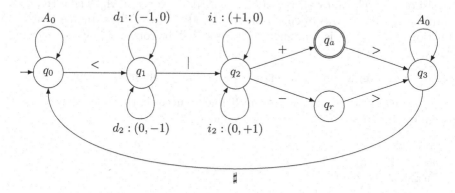

Fig. 1. The VASS $\mathcal{A}_2$ with two counters that recognises a Σ_1^1-complete ω-language

Lemma 2. *There exists a continuous reduction from* $\mathrm{IF}_{\mathrm{pre}}$ *to the ω-language recognised by* $\mathcal{A}_2$.

Intuition. An ω-word accepted by $\mathcal{A}_2$ consists of infinitely many *phases* separated by $\sharp$. Each phase is a finite word over the alphabet A_0. In our reduction we will restrict to phases being sequences of *blocks*, each block being a finite word of the form given by the following definition (for $n_1, n_2, m_1, m_2 \in \omega$ and $s \in \{+, -\}$):

$$B^s(-n_1, -n_2, +m_1, +m_2) \stackrel{\text{def}}{=} \; < d_1^{m_1} \, d_2^{n_2} \mid i_1^{m_1} \, i_2^{m_2} \, s > \; \in A_0^*. \qquad (3)$$

Such a block is *accepting* if $s = +$, otherwise $s = -$ and the block is *rejecting*. If $\mathcal{A}_2$ starts reading a block and moves from q_0 to q_1 over $<$ then we say that it *chooses* this block. Otherwise $\mathcal{A}_2$ stays in q_0 and it does not *choose* the given block. By the construction of the VASS $\mathcal{A}_2$, in every run it needs to choose exactly one block from each phase. Additionally, the run is accepting if and only if infinitely many of the chosen blocks are accepting.

In our reduction we will represent a given set $X \subseteq \mathcal{T}$ by an appropriately defined sequence of phases. We will control the set of configurations the VASS can reach at the beginning of each phase. These configurations will form an antichain with respect to the coordinate-wise (or simulation) order: if the VASS can reach two distinct configurations (q_0, c_1, c_2) and (q_0, c_1', c_2') then either $c_1 < c_1'$ and $c_2 > c_2'$; or $c_1 > c_1'$ and $c_2 < c_2'$. Each block in the successive phase will be of the form $B^s(-c_1, -c_2, +m_1, +m_2)$ for some reachable configuration (q_0, c_1, c_2)—this will be the only reachable configuration in which the automaton can choose the considered block. After choosing it, the automaton will finish reading the phase in the configuration (q_3, m_1, m_2).

Proof of Lemma 2. For the rest of this section we prove Lemma 2. Let us fix a set $X \subseteq \mathcal{T}$. We will construct an ω-word $\alpha(X) \in A^\omega$. The ω-word $\alpha(X)$ will consist of infinitely many phases $\alpha(X) = u_0 \sharp u_1 \sharp \cdots$, for $u_n \in A_0^*$. The n-th phase u_n (for $n = 0, 1, \ldots$) will depend on $X \cap \{\text{L}, \text{R}\}^n$. This will guarantee that the function $\alpha \colon 2^{\mathcal{T}} \to A^\omega$ is continuous. The proof will be concluded by the following claim.

Claim. X has an infinite $\preceq$-ascending chain if and only if $\mathcal{A}_2$ accepts $\alpha(X)$.

To simplify the construction, let us define inductively the function $b \colon \mathcal{T} \to \omega$, assigning to nodes $v \in \mathcal{T}$ their binary value $b(v)$:

- $b(\epsilon) = 0$,
- $b(v\text{L}) = 2 \cdot b(v)$,
- $b(v\text{R}) = 2 \cdot b(v) + 1$.

Let $b'(v) = 2^n - b(v) - 1$ for $n = |v|$ (i.e. $v \in \{\text{L}, \text{R}\}^n$). Note that for every $n \in \omega$ we have

$$b(\{\text{L}, \text{R}\}^n) = b'(\{\text{L}, \text{R}\}^n) = \{0, 1, \ldots, 2^n - 1\},$$

and both b and b' are bijective between these sets. Additionally, if $v \neq v' \in \{\text{L}, \text{R}\}^n$ then either $b(v) < b(v')$ and $b'(v) > b'(v')$; or $b(v) > b(v')$ and $b'(v) < b'(v')$.

We take any $n = 0, 1, \ldots$ and define the n-th phase u_n. Let u_n be the concatenation of the following blocks, for all $v \in \{\text{L}, \text{R}\}^n$ and $d \in \{\text{L}, \text{R}\}$:

$$B^s\big(-b(v), -b'(v), +b(vd), +b'(vd)\big),$$

where $s = +$ if $v \in X$ and $s = -$ otherwise. Thus, the n-th phase is a concatenation of 2^{n+1} blocks, one for each node vd in $\{\text{L}, \text{R}\}^{n+1}$.

To prove Claim 2 it is enough to notice the following fact.

Fact 4. *There is a bijection between infinite branches $\beta \in \{\text{L}, \text{R}\}^\omega$ and runs ρ of $\mathcal{A}_2$ over $\alpha(X)$. The bijection satisfies that the configuration in ρ before reading the n-th phase of $\alpha(X)$ is $\big(q_0, b(v_n), b'(v_n)\big)$ for $v_n = \beta\restriction_n \in \{\text{L}, \text{R}\}^n$. $\mathcal{A}_2$ visits an accepting state in ρ while reading the n-th phase of $\alpha(X)$ if and only if $v_n \in X$.*

Proof. Easy induction. $\square$

This concludes the proof of Lemma 2.

3   Representation of IF_{pre}

To construct our continuous reduction in the one-counter case, we need the following simple lemma that provides an alternative characterisation of the set IF_{inf}. Let us introduce the following definition.

Definition 1. *A sequence $v_0, v_1 \ldots \in \mathcal{T}$ is called a* correct chain *if $v_0 = \epsilon$ and for every $n = 0, 1, \ldots$:*

1. *$|v_{n+1}| = |v_n| + 1$,*
2. *$v_{n+1} \leq_{\inf} v_n \text{R}$ (or equivalently $v_{n+1} \leq_{\text{lex}} v_n \text{R}$).*

A correct chain is witnessing *for a set $X \subseteq \mathcal{T}$ if for infinitely many n we have $v_n \in X$ and $v_{n+1} \leq_{\inf} v_n \text{L}$.*

Intuitively, the definition forces the sequence to be *not so-much increasing* in the infix order $\leq_{\inf}$: the successive element v_{n+1} needs to be *to the left* in the tree from $v_n \text{R}$. Such a sequence is *witnessing* for a set X if infinitely many times it belongs to X and at these moments it actually drops in $\leq_{\inf}$.

Lemma 3. *A set $X \subseteq \mathcal{T}$ belongs to IF_{inf} if and only if there exists a correct chain witnessing for X.*

Proof. First take a correct chain witnessing for X. Let $x_0, x_1, \ldots$ be the subsequence that shows that $(v_n)_{n \in \omega}$ is witnessing for X. In that case, by the definition, for all n we have $x_n \in X$ and $x_{n+1} <_{\inf} x_n$ (because $x_{n+1} \text{M}^\omega \leq_{\text{lex}} x_n \text{LR}^\omega <_{\text{lex}} x_n \text{M}^\omega$). Thus, X has an infinite $\leq_{\inf}$-descending chain and belongs to IF_{inf}.

Now assume that $X \in \text{IF}_{\text{inf}}$ and $x_0 >_{\inf} x_1 >_{\inf} x_2 >_{\inf} \ldots$ is a sequence witnessing that. Without loss of generality we can assume that $|x_{n+1}| > |x_n|$ because for each fixed depth k there are only finitely many nodes of T in $\{\text{L},\text{R}\}^{\leq k}$. We can now add intermediate nodes in-between the sequence $(x_n)_{n\in\omega}$ to construct a correct chain witnessing for X; the following pseudo-code realises this goal:

```
n := 0;
i := 0;
while (true) {
  if (n > |x_i|) {
    i := i+1;
  }
  v_n := x_i⌈n;
  n := n+1;
}
```

Clearly, Property 1 in the definition of a correct chain is guaranteed. Let $i \in \omega$ and $n = |x_i|$. By the fact that $x_{i+1} <_{\text{inf}} x_i$ we know that $x_{i+1}\!\restriction_{n+1} \leq_{\text{inf}} x_i\text{L}$. Therefore, for every $n \in \omega$ we have $v_{n+1} \leq_{\text{inf}} v_n\text{R}$ and if $n = |x_i|$ for some i then $v_{n+1} \leq_{\text{inf}} v_n\text{L}$. It implies that the sequence $(v_n)_{n\in\omega}$ satisfies Property 1 in the definition of a correct chain. It is clearly witnessing for X because it contains $(x_n)_{n\in\omega}$ as a subsequence.   □

4   Hardness for One Counter

In this section we provide an example of an ω-language that is Σ_1^1-complete and can be recognised by a VASS $\mathcal{A}_1$ with one counter. $\mathcal{A}_1$ is depicted in Fig. 2, it is very similar to $\mathcal{A}_2$, but simpler. Let $A_0 \overset{\text{def}}{=} \{<, d, |, i, +, -, >\}$ and let the alphabet $A \overset{\text{def}}{=} A_0 \cup \{\sharp\}$.

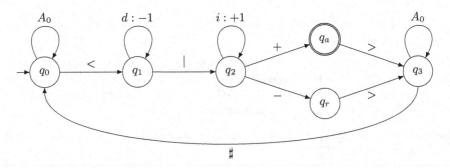

Fig. 2. The VASS $\mathcal{A}_1$ with one counter that recognises a Σ_1^1-complete ω-language

Proposition 1. *There exists a continuous reduction from* IF_{inf} *to the* ω-*language recognised by* $\mathcal{A}_1$.

Similarly as before, we will use the notion of phases and blocks. Since there is only one counter now (and only two letters modifying its value: d and i) we exchange the definition of a block (see (3)) by the following one (for $n, m \in \omega$ and $s \in \{+, -\}$):

$$B^s(-n, +m) \stackrel{\text{def}}{=} \; < d^n \mid i^m \; s > \; \in A_0^*. \tag{4}$$

Similarly as before, we will take a set $X \subseteq T$ and construct an ω-word $\alpha(X)$. This ω-word will be a concatenation of infinitely many phases $u_0 \sharp u_1 \sharp \cdots$. The n-th phase u_n will depend on $X \cap \{\text{L}, \text{R}\}^n$. The configurations (q_0, c) reached at the beginning of an n-th phase will be in correspondence with the nodes $v \in \{\text{L}, \text{R}\}^n$. The bigger the value c, the higher in the infix order (or the lexicographic order, as they overlap here) the respective node v is.

To precisely define our ω-word $\alpha(X)$ we need to define fast-growing functions: $m \colon \{-1\} \cup \omega \to \omega$ and $e \colon T \to \omega$:

$$
\begin{aligned}
m(-1) &= 1, \\
m(n) &= m(n-1) \cdot 2^n && \text{for } n \in \omega, \\
e(v) &= m(|v| - 1) \cdot b(v) && \text{for } v \in T.
\end{aligned}
$$

These functions allow to use a big range of the possible values of a single counter of a VASS to represent particular nodes of the tree. We will use the following two invariants of this definition, for $n \in \omega$ and $v, v' \in \{\text{L}, \text{R}\}^n$:

$$v <_{\inf} v' \iff e(v) \le e(v'), \tag{5}$$
$$e(v) + m(|v| - 1) \; \le \; m(|v|). \tag{6}$$

We take any $n = 0, 1, \ldots$ and define the n-th phase u_n. Let u_n be the concatenation of the following blocks, for all $v \in \{\text{L}, \text{R}\}^n$ and $d \in \{\text{L}, \text{R}\}$:

$$B^s\big(-e(v), +e(vd)\big),$$

where $s = +$ if $v \in X$ and $d = \text{L}$; otherwise $s = -$. Thus, the n-th phase is a concatenation of 2^{n+1} blocks, one for each node vd in $\{\text{L}, \text{R}\}^{n+1}$.

To conclude the proof of Proposition 1 it is enough to prove the following two lemmas.

Lemma 4. *If there exists a correct chain witnessing for X then $\alpha(X) \in \text{L}(\mathcal{A}_1)$.*

Lemma 5. *If $\alpha(X) \in \text{L}(\mathcal{A}_1)$ then there exists a correct chain witnessing for X.*

Proof of Lemma 4. Consider a correct chain $(v_n)_{n \in \omega}$ witnessing for X. Assume that $I \subseteq \omega$ is an infinite set such that for $n \in I$ we have $v_n \in X$ and $v_{n+1} \le_{\inf} v_n \text{L}$. Let us construct inductively a run ρ of $\mathcal{A}_1$ on $\alpha(X)$. The invariant is that for each $n \in \omega$ the configuration of ρ before reading the n-th phase of $\alpha(X)$ is of the form (q_0, c_n) with $c_n \ge e(v_n)$. To define ρ it is enough to decide which block to choose from an n-th phase of $\alpha(X)$:

– if $n \in I$ then choose the block $B^+\big(-e(v_n), +e(v_{n\text{L}})\big)$,

– otherwise choose the block $B^-\big(-e(v_n), +e(v_{n\text{R}})\big)$.

Notice that by the invariant, it is allowed to choose the respective blocks as $c_n \geq e(v_n)$. Because of (5) and the fact that $(v_n)_{n\in\omega}$ is a correct chain, the invariant is preserved. As the set I is infinite, the constructed run chooses an accepting block infinitely many times and thus is accepting. □

Proof of Lemma 5. Assume that ρ is an accepting run of $\mathcal{A}_1$ over $\alpha(X)$. For $n = 0, 1, \ldots$ let (q_0, c_n) be the configuration in ρ before reading the n-th phase of $\alpha(X)$ and assume that ρ chooses a block of the form $B^{s_n}\big(-e(v_n), +e(v_n d_n)\big)$ in the n-th phase of $\alpha(X)$. Our aim is to show that $(v_n)_{n\in\omega}$ is a correct chain witnessing for X. First notice that by the construction of $\alpha(X)$ we have $|v_n| = n$.

Clearly, as the counter needs to be non-negative, we have $e(v_n) \leq c_n$. Notice that by (6) we obtain inductively for $n = 0, 1, \ldots$ that $c_n < m(n)$. Therefore, we have

$$
\begin{aligned}
m(n) \cdot b(v_{n+1}) = e(v_{n+1}) \leq c_{n+1} \\
= c_n - e(v_n) + e(v_n d_n) < m(n) + e(v_n d_n) \\
= m(n) + m(n) \cdot b(v_n d_n).
\end{aligned}
$$

By dividing by $m(n)$ we obtain $b(v_{n+1}) < 1 + b(v_n d_n)$, thus $b(v_{n+1}) \leq b(v_n d_n)$ and therefore $v_{n+1} \leq_{\inf} v_n d_n \leq_{\inf} v_{n\text{R}}$. Moreover, if $s_n = +$ (i.e. the n-th chosen block is accepting) then $v_n \in X$ and $d_n = \text{L}$. Therefore, as ρ chooses infinitely many accepting blocks, $(v_n)_{n\in\omega}$ is witnessing for X. □

This concludes the proof of Proposition 1.

5   Inherent Non-determinism

In this section we formally state and prove Corollary 1. It is expressed in the same spirit as the corresponding Theorem 5.5 in [11]: we consider an abstract model of automata $\mathcal{A}$ with a countable set of configurations C, an initial configuration $c_I \in C$, a transition relation $\delta \subseteq C \times A \times C$, and an acceptance condition $W \subseteq C^\omega$. The notions of a run $\text{run}(\alpha, \rho)$; an accepting run $\text{acc}(\rho)$; and the language $\text{L}(\mathcal{A})$ are defined in the standard way. Thus, under the assumption that the acceptance condition W is Borel, the set

$$
P \overset{\text{def}}{=} \big\{(\alpha, \rho) \in A^\omega \times C^\omega \mid \text{run}(\alpha, \rho) \wedge \text{acc}(\rho)\big\},
$$

as in (1) is also Borel. The assumptions that the machine is deterministic, unambiguous, or countably-unambiguous imply that the cardinality of the sections $P_\alpha \overset{\text{def}}{=} \{\rho \mid (\alpha, \rho) \in P\}$ for $\alpha \in A^\omega$ is at most countable. Therefore, the following *small section theorem* by Lusin and Novikov applies.

Theorem 2 (see [12, Theorem 18.10]). *Let X, Y be standard Borel spaces and let $P \subseteq X \times Y$ be Borel. If every section P_x is countable, then P has a Borel uniformization and therefore $\pi_X(P)$ is Borel.*

Therefore, we know that $L(\mathcal{A}) = \pi_{A^\omega}(P)$ is Borel. Thus, no such machine can recognise $L(\mathcal{A}_1)$ for the Büchi VASS $\mathcal{A}_1$ from Sect. 4, as that language is non-Borel.

6   Concluding Remarks

The core result of this paper is a technique of encoding a Σ^1_1-complete set in a *monotone* way using only one partially blind counter—Proposition 1. This shows that even in that restricted case, the non-determinism of the machines is inherent, and cannot be simulated by any restricted form (like countable-unambiguity).

The question whether one counter Büchi VASS recognise languages at all levels of the Wadge hierarchy that are occupied by non-deterministic Büchi Turing machines (see [9]) is left open. The construction provided in [9] involves four counters and at the moment it is not clear whether one can reduce that number.

References

1. Bojańczyk, M.: Weak MSO with the unbounding quantifier. Theory Comput. Syst. **48**(3), 554–576 (2011)
2. Bojańczyk, M., Colcombet, T.: Bounds in ω-regularity. In: LICS, pp. 285–296 (2006)
3. Bojańczyk, M., Toruńczyk, S.: Deterministic automata and extensions of weak MSO. In: FSTTCS, pp. 73–84 (2009)
4. Cabessa, J., Duparc, J., Facchini, A., Murlak, F.: The Wadge hierarchy of max-regular languages. In: FSTTCS, pp. 121–132 (2009)
5. Colcombet, T.: Forms of determinism for automata. In: STACS, pp. 1–23 (2012)
6. Duparc, J., Finkel, O., Ressayre, J.P.: Computer science and the fine structure of Borel sets. Theor. Comput. Sci. **257**(1–2), 85–105 (2001)
7. Duparc, J., Finkel, O., Ressayre, J.-P.: The Wadge hierarchy of petri nets ω-languages. In: Artemov, S., Nerode, A. (eds.) LFCS 2013. LNCS, vol. 7734, pp. 179–193. Springer, Heidelberg (2013). https://doi.org/10.1007/978-3-642-35722-0_13
8. Finkel, O.: Borel ranks and Wadge degrees of context free omega-languages. Math. Struct. Comput. Sci. **16**(5), 813–840 (2006)
9. Finkel, O.: Wadge degrees of ω-languages of Petri nets (2018)
10. Finkel, O., Skrzypczak, M.: On the topological complexity of w-languages of non-deterministic Petri nets. Inf. Process. Lett. **114**(5), 229–233 (2014)
11. Hummel, S., Skrzypczak, M.: The topological complexity of MSO+U and related automata models. Fundam. Inform. **119**(1), 87–111 (2012)
12. Kechris, A.: Classical Descriptive Set Theory. Springer, New York (1995). https://doi.org/10.1007/978-1-4612-4190-4
13. Thomas, W., Lescow, H.: Logical specifications of infinite computations. In: de Bakker, J.W., de Roever, W.-P., Rozenberg, G. (eds.) REX 1993. LNCS, vol. 803, pp. 583–621. Springer, Heidelberg (1994). https://doi.org/10.1007/3-540-58043-3_29

Qualitative Reachability for Open Interval Markov Chains

Jeremy Sproston[✉]

Dipartimento di Informatica, University of Turin, Turin, Italy
sproston@di.unito.it

Abstract. Interval Markov chains extend classical Markov chains with the possibility to describe transition probabilities using intervals, rather than exact values. While the standard formulation of interval Markov chains features closed intervals, previous work has considered also open interval Markov chains, in which the intervals can also be open or half-open. In this paper we focus on qualitative reachability problems for open interval Markov chains, which consider whether the optimal (maximum or minimum) probability with which a certain set of states can be reached is equal to 0 or 1. We present polynomial-time algorithms for these problems for both of the standard semantics of interval Markov chains. Our methods do not rely on the closure of open intervals, in contrast to previous approaches for open interval Markov chains, and can characterise situations in which probability 0 or 1 can be attained not exactly but arbitrarily closely.

1 Introduction

The development of modern computer systems can benefit substantially from a verification phase, in which a formal model of the system is exhaustively verified in order to identify undesirable errors or inefficiencies. In this paper we consider the verification of probabilistic systems, in which state-to-state transitions are accompanied by probabilities that specify the relative likelihood with which the transitions occur, using model-checking techniques; see [1,2,11] for general overviews of this field. One drawback of classical formalisms for probabilistic systems is that they typically require the specification of exact probability values for transitions: in practice, it is likely that such precise information concerning the probability of system behaviour is not available. A solution to this problem is to associate intervals of probabilities with transitions, rather than exact probability values, leading to *interval Markov chains* (IMCs) or *interval Markov decision processes*. IMCs have been studied in [14,15], and considered in the *qualitative* and *quantitative* model-checking context in [5,6,18]. Qualitative model checking concerns whether a property is satisfied by the system model with probability (equal to or strictly greater than) 0 or (equal to or strictly less than) 1, whereas quantitative model checking considers whether a property is satisfied with probability above or below some threshold in the interval $[0,1]$, and

© Springer Nature Switzerland AG 2018
I. Potapov and P.-A. Reynier (Eds.): RP 2018, LNCS 11123, pp. 146–160, 2018.
https://doi.org/10.1007/978-3-030-00250-3_11

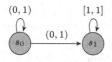

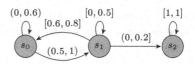

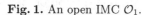

Fig. 1. An open IMC $\mathcal{O}_1$.   **Fig. 2.** An open IMC $\mathcal{O}_2$.

generally involves the computation of the probability of property satisfaction, which is then compared to the aforementioned threshold.

In [5,6,18], the intervals associated with transitions are *closed*. This limitation was adressed in [4], which considered the possibility of utilising *open* (and half-open) intervals, in addition to closed intervals. Example of such open IMCs are shown in Figs. 1 and 2. In [4], it was shown that the probability of the satisfaction of a property in an open IMC can be approximated arbitrarily closely by a standard, closed IMC obtained by changing all (half-)open intervals featured in the model to closed intervals with the same endpoints. However, although the issue of the determining the existence of exact solutions is mentioned in [4], closing the intervals can involve the loss of information concerning exact solutions. Take, for example, the open IMC in Fig. 1: changing the intervals from $(0,1)$ to $[0,1]$ on both of the transitions means that the minimum probability of reaching the state s_1 after starting in state s_0 becomes 0, whereas the probability of reaching s_1 from s_0 is strictly greater than 0 for all ways of assigning probabilities to the transitions in the original IMC.

In this paper we propose verification methods for qualitative reachability properties of open IMCs. We consider both of the standard semantics for IMCs. The uncertain Markov chain (UMC) semantics associated with an IMC comprises an infinite number of standard Markov chains, each corresponding to a certain choice of probability for each transition. In contrast, the interval Markov decision process (IMDP) semantics associates a single Markov decision process (MDP) with the IMC, where from each state there is available an uncountable number of distributions, each corresponding to one assignment of probabilities belonging to the intervals of the transitions leaving that state. The key difference between the two semantics can be summarised by considering the behaviour from a particular state of the IMC: in the UMC semantics, the same probability distribution over outgoing transitions must always be used from the state, whereas in the IMDP semantics the outgoing probability distribution may change for each visit to the state. We show that we can obtain exact (not approximate) solutions for both semantics in polynomial time in the size of the open IMC.

For the UMC semantics, and for three of the four classes of qualitative reachability problem in the IMDP semantics, the algorithms presented are inspired by methods for finite MDPs. In the case of the IMDP semantics, these algorithms rely on the fact that retaining the memory of previous choices along the behaviour of an IMC is not necessary. A direct method for the construction of a finite MDP that represents an IMC and which can be used for the verification of qualitative properties is the following: the set of states of the finite MDP equals that of the IMC and, for each state s and each set X of states, there exists a single distribution from s in the finite MDP that assigns positive probability to each state in X if and only if there exists at least one probability assignment for transitions in the IMC that assigns positive probability to each transition from s with target state in X. Intuitively, a distribution associated with s and X in the finite MDP can be regarded as the representative distribution of all probability assignments of the IMC that assign positive probability to the transitions from s to states in X. However, such a finite MDP construction does not yield polynomial-time algorithms in the size of the open IMC, because the presence of transitions having zero as their left endpoint can result in an exponential number of distributions in the number of IMC transitions. In our methods, apart from considering issues concerning the difference between closed and open intervals and the subsequent implications for qualitative reachability problems, we avoid such an exponential blow up. In particular, we show how the predecessor operations used by some qualitative reachability algorithms for MDPs can be applied directly on the open IMC.

The fourth class of reachability problem in the IMDP semantics concerns determining whether the probability of reaching a certain set of states from the current state is equal to 1 for all schedulers, where a scheduler chooses an outgoing probability distribution from a state on the basis of the choices made so far. For this class of problem, retaining memory of previous choices can be important for showing that the problem is *not* satisfied, i.e., that there exists a scheduler such that the reachability probability is strictly less than 1. As an example, we can take the open IMC in Fig. 1. Consider the memoryful scheduler that assigns probability $\frac{1}{2^i}$ to the i-th attempt to take a transition from s_0 to s_1, meaning that the overall probability of reaching s_1 when starting in s_0 under this scheduler is $\frac{1}{2} + \frac{1}{2}(\frac{1}{4} + \frac{3}{4}(\frac{1}{8} + \cdots)) < 1$. Instead a memoryless scheduler will reach s_1 with probability 1: for any $\lambda \in (0, 1)$ representing the (constant) probability of taking the transition from s_0 to s_1, the overall probability of reaching s_1 is $\lim_{k \to \infty} 1 - (1 - \lambda)^k = 1$. Hence our results for this class of reachability problem take the inadequacy of memoryless schedulers into account; indeed, while the algorithms presented for all other classes of problems (and all problems for the UMC semantics) proceed in a manner similar to that introduced in the literature for finite MDPs, for this class we present an *ad hoc* algorithm, based on an adaptation of the classical notion of end components [9].

After introducing open IMCs in Sect. 2, the algorithms for the UMC semantics and the IMDP semantics are presented in Sects. 3 and 4, respectively. The proofs of the results can be found in [20].

Related Work. Model checking of qualitative properties of Markov chains (see, for example, [7,21]) relies on the fact that transition probability values are fixed throughout the behaviour of the system, and does not require that exact probability values are taken into account during analysis. The majority of work on model checking for IMCs considers the more general quantitative problems: [5,18] present algorithms utilising a finite MDP construction based on encoding within distributions available from a state the extremal probabilities allowed from that state (known as the state's basic feasible solutions). Such a construction results in an exponential blow up, which is also not avoided in [5] for qualitative properties (when transitions can have 0 as their left endpoint). [6,17] improve on these results to present polynomial-time algorithms for reachability problems based on linear or convex programming. The paper [12] includes polynomial-time methods for computing (maximal) end components, and for computing a single step of value iteration, for interval MDPs. We note that IMCs are a special case of constraint Markov chains [3], and that the UMC semantics of IMCs corresponds to a special case of parametric Markov chains [8,16]. As far as we are aware, only [4] considers open IMCs.

2   Open Interval Markov Chains

Preliminaries. A *(probability) distribution* over a finite set Q is a function $\mu : Q \to [0,1]$ such that $\sum_{q \in Q} \mu(q) = 1$. Let $\mathsf{Dist}(Q)$ be the set of distributions over Q. We use $\mathsf{support}(\mu) = \{q \in Q \mid \mu(q) > 0\}$ to denote the *support set* of μ, i.e., the set of elements assigned positive probability by μ, and use $\{q \mapsto 1\}$ to denote the distribution that assigns probability 1 to the single element q. Given a binary function $f : Q \times Q \to [0,1]$ and element $q \in Q$, we denote by $f(q, \cdot) : Q \to [0,1]$ the unary function such that $f(q, \cdot)(q') = f(q, q')$ for each $q' \in Q$. We let $\mathcal{I}$ denote the set of (open, half-open or closed) intervals that are subsets of $[0,1]$ and that have rational-numbered endpoints. Given an interval $I \in \mathcal{I}$, we let $\mathsf{left}(I)$ (respectively, $\mathsf{right}(I)$) be the left (respectively, right) endpoint of I. The set of closed (respectively, left-open, right-closed; left-closed, right-open; open) intervals in $\mathcal{I}$ is denoted by $\mathcal{I}^{[\cdot,\cdot]}$ (respectively, $\mathcal{I}^{(\cdot,\cdot]}$; $\mathcal{I}^{[\cdot,\cdot)}$; $\mathcal{I}^{(\cdot,\cdot)}$). Hence we have $\mathcal{I} = \mathcal{I}^{[\cdot,\cdot]} \cup \mathcal{I}^{(\cdot,\cdot]} \cup \mathcal{I}^{[\cdot,\cdot)} \cup \mathcal{I}^{(\cdot,\cdot)}$. Furthermore, we let $\mathcal{I}^{(+,\cdot)}$ (respectively, $\mathcal{I}^{[0,\cdot)}$; $\mathcal{I}^{(0,\cdot)}$) be the set of intervals in $\mathcal{I}$ such that the left endpoint is positive (respectively, left-closed intervals with the left endpoint equal to 0; left-open intervals with the left endpoint equal to 0). Finally, let $\mathcal{I}^{(0,\cdot)} = \mathcal{I}^{[0,\cdot)} \cup \mathcal{I}^{(0,\cdot)}$ be the set of intervals in $\mathcal{I}$ with left endpoint equal to zero.

A *discrete-time Markov chain* (DTMC) $\mathcal{D}$ is a pair $(S, \mathbf{P})$ where S is a set of *states*, and $\mathbf{P} : S \times S \to [0,1]$ is a *transition probability matrix*, such that, for each state $s \in S$, we have $\sum_{s' \in S} \mathbf{P}(s, s') = 1$. Note that $\mathbf{P}(s, \cdot)$ is a distribution, for each state $s \in S$. A *path* of DTMC $\mathcal{D}$ is a sequence $s_0 s_1 \cdots$ such that $\mathbf{P}(s_i, s_{i+1}) > 0$ for all $i \geq 0$. Given a path $\rho = s_0 s_1 \cdots$ and $i \geq 0$, we let $\rho(i) = s_i$ be the $(i+1)$-th state along ρ. The set of paths of $\mathcal{D}$ starting in state $s \in S$ is denoted by $Paths^{\mathcal{D}}(s)$. In the standard manner (see, for example, [2,11]), given a state $s \in S$, we can define a probability measure $\Pr_s^{\mathcal{D}}$ over $Paths^{\mathcal{D}}(s)$.

A *Markov decision process* (MDP) $\mathcal{M}$ is a pair (S, Δ) where S is a finite set of *states* and $\Delta : S \to 2^{\mathsf{Dist}(S)}$ is a *transition function* such that $\Delta(s) \neq \emptyset$ for all $s \in S$. We say that an MDP is *finite* if $\Delta(s)$ is finite for all $s \in S$.

A(n infinite) path of an MDP $\mathcal{M}$ is a sequence $s_0 \mu_0 s_1 \mu_1 \cdots$ such that $\mu_i \in \Delta(s_i)$ and $\mu_i(s_{i+1}) > 0$ for all $i \geq 0$. Given a path $\rho = s_0 \mu_0 s_1 \mu_1 \cdots$ and $i \geq 0$, we let $\rho(i) = s_i$ be the $(i+1)$-th state along ρ. A finite path is a sequence $r = s_0 \mu_0 s_1 \mu_1 \cdots \mu_{\mu_{n-1}} s_n$ such that $\mu_i \in \Delta(s_i)$ and $\mu_i(s_{i+1}) > 0$ for each $0 \leq i < n$. Let $last(r) = s_n$ denote the final state of r. Let $Paths_*^{\mathcal{M}}$ be the set of finite paths of the MDP $\mathcal{M}$. Let $Paths^{\mathcal{M}}(s)$ and $Paths_*^{\mathcal{M}}(s)$ be the sets of infinite paths and finite paths, respectively, of $\mathcal{M}$ starting in state $s \in S$.

A *scheduler* is a mapping $\sigma : Paths_*^{\mathcal{M}} \to \mathsf{Dist}(\bigcup_{s \in S} \Delta(s))$ such that $\sigma(r) \in \mathsf{Dist}(\Delta(last(r)))$ for each $r \in Paths_*^{\mathcal{M}}$. Let $\Sigma^{\mathcal{M}}$ be the set of schedulers of the MDP $\mathcal{M}$. Given a state $s \in S$ and a scheduler σ, we can define a countably infinite-state DTMC $\mathcal{D}_s^\sigma$ that corresponds to the behaviour of the scheduler σ from state s, which in turn can be used to define a probability measure Pr_s^σ over $Paths^{\mathcal{M}}(s)$ in the standard manner (see [2,11]). A scheduler $\sigma \in \Sigma^{\mathcal{M}}$ is *memoryless* if, for finite paths $r, r' \in Paths_*^{\mathcal{M}}$ such that $last(r) = last(r')$, we have $\sigma(r) = \sigma(r')$. Let $\Sigma_{\mathrm{m}}^{\mathcal{M}}$ be the set of memoryless schedulers of $\mathcal{M}$. Note that, for a memoryless scheduler $\sigma \in \Sigma_{\mathrm{m}}^{\mathcal{M}}$, we can construct a finite DTMC $\tilde{\mathcal{D}}^\sigma = (S, \tilde{\mathbf{P}})$ with $\tilde{\mathbf{P}}(s, s') = \sum_{\mu \in \Delta(s)} \sigma(s)(\mu) \cdot \mu(s')$: we call this DTMC the *folded DTMC of σ*. It can be shown that Pr_s^σ and $\mathrm{Pr}_s^{\tilde{\mathcal{D}}^\sigma}$ assign the same probabilities to measurable sets of paths, because the state s of the DTMC $\mathcal{D}_s^\sigma$ is probabilistic bisimulation equivalent to the state s of the folded DTMC $\tilde{\mathcal{D}}^\sigma$ (for a definition of probabilistic bisimulation and more information on this point, see [2, Sect. 10.4.2]).

Interval Markov Chains: Syntax. An (open) *interval Markov chain* (IMC) $\mathcal{O}$ is a pair (S, δ), where S is a finite set of *states*, and $\delta : S \times S \to \mathcal{I}$ is a *interval-based transition function*.

In the following, we refer to *edges* as those state pairs for which the transition function does not assign the probability 0 point interval $[0,0]$. Formally, let the set of edges E of $\mathcal{O}$ be defined as $\{(s, s') \in S \times S \mid \delta(s, s') \neq [0,0]\}$. We use edges to define the notion of path for IMCs: a path of an IMC $\mathcal{O}$ is a sequence $s_0 s_1 \cdots$ such that $(s_i, s_{i+1}) \in E$ for all $i \geq 0$. Given a path $\rho = s_0 s_1 \cdots$ and $i \geq 0$, we let $\rho(i) = s_i$ be the $(i+1)$-th state along ρ. We use $Paths^{\mathcal{O}}$ to denote the set of paths of $\mathcal{O}$, $Paths_*^{\mathcal{O}}$ to denote the set of finite paths of $\mathcal{O}$, and $Paths^{\mathcal{O}}(s)$ and $Paths_*^{\mathcal{O}}(s)$ to denote the sets of paths and finite paths starting in state $s \in S$.

Given a state $s \in S$, we say that a distribution $\mathbf{a} \in \mathsf{Dist}(S)$ is an *assignment for s* if $\mathbf{a}(s') \in \delta(s, s')$ for each state $s' \in S$. We say that the IMC $\mathcal{O}$ is *well formed* if there exists at least one assignment for each state. Note that an assignment for state $s \in S$ exists if and only if the following conditions hold: (1a) $\sum_{s' \in S} \mathsf{left}(\delta(s, s')) \leq 1$, (1b) $\sum_{s' \in S} \mathsf{left}(\delta(s, s')) = 1$ implies that $\delta(s, s')$ is left-closed for all $s' \in S$, (2a) $\sum_{s' \in S} \mathsf{right}(\delta(s, s')) \geq 1$, and (2b) $\sum_{s' \in S} \mathsf{right}(\delta(s, s')) = 1$ implies that $\delta(s, s')$ is right-closed for all $s' \in S$. We henceforth consider IMCs that are well formed. We define the *size of an IMC* $\mathcal{O} = (S, \delta)$ as the size of the representation of δ, which is the sum over all states $s, s' \in S$ of the binary representation of the endpoints of $\delta(s, s')$, where rational numbers are encoded as the quotient of integers written in binary.

Interval Markov Chains: Semantics. IMCs are typically presented with regard to two semantics, which we consider in turn. Given an IMC $\mathcal{O} = (S, \delta)$, the *uncertain Markov chain* (UMC) semantics of $\mathcal{O}$, denoted by $[\mathcal{O}]_{\mathrm{U}}$, is the smallest set of DTMCs such that $(S, \mathbf{P}) \in [\mathcal{O}]_{\mathrm{U}}$ if, for each state $s \in S$, the distribution $\mathbf{P}(s, \cdot)$ is an assignment for s. The *interval Markov decision process* (IMDP) semantics of $\mathcal{O}$, denoted by $[\mathcal{O}]_{\mathrm{I}}$, is the MDP (S, Δ) where, for each state $s \in S$, we let $\Delta(s)$ be the set of assignments for s.

Reachability. Let $\mathcal{O} = (S, \delta)$ be an IMC and let $T \subseteq S$ be a set of states. We define $\mathsf{Reach}(T) \subseteq Paths^{\mathcal{O}}$ to be the set of paths of $\mathcal{O}$ that reach at least one state in T. Formally, $\mathsf{Reach}(T) = \{\rho \in Paths^{\mathcal{O}} \mid \exists i \in \mathbb{N}.\rho(i) \in T\}$. In the following we assume without loss of generality that states in T are absorbing in all the IMCs that we consider, i.e., $\delta(s, s) = [1, 1]$ for all states $s \in T$.

Edge Sets. Let $\mathcal{O} = (S, \delta)$ be an IMC. Let $s \in S$ be a state of $\mathcal{O}$, and let $E(s) = \{(s, s') \in E \mid s' \in S\}$ be the set of edges of $\mathcal{O}$ with source s. Let $\star \in \{[\cdot, \cdot], (\cdot, \cdot], [\cdot, \cdot), (\cdot, \cdot), \langle +, \cdot \rangle, [0, \cdot \rangle, (0, \cdot \rangle, \langle 0, \cdot \rangle\}$, and let $E^{\star} = \{(s, s') \in E \mid \delta(s, s') \in \mathcal{I}^{\star}\}$. Given $X \subseteq S$, and given s and $\star$ as defined above, let $E(s, X) = \{(s, s') \in E(s) \mid s' \in X\}$ and $E^{\star}(s, X) = E(s, X) \cap E^{\star}$.

Valid Edge Sets. We are interested in identifying the sets of edges from state $s \in S$ that result from assignments. Such a set is characterised by two syntactic conditions: the first condition requires that the sum of the upper bounds of the set's edges' intervals is at least 1, whereas the second condition specifies the edges from state s that are *not* included in the set can be assigned probability 0. Formally, we say that a non-empty subset $B \subseteq E(s)$ of edges from s is *large* if either (a) $\sum_{e \in B} \mathsf{right}(\delta(e)) > 1$ or (b) $\sum_{e \in B} \mathsf{right}(\delta(e)) = 1$ and $B \subseteq E^{\langle \cdot, \cdot]}$. The set B is *realisable* if $E(s) \backslash B \subseteq E^{[0, \cdot)}$. Then we say that $B \subseteq E(s)$ is *valid* if it is large and realisable. The following lemma specifies that a valid edge set for state s characterises exactly the support sets of some assignments for s.

Lemma 1. *Let $s \in S$ and $B \subseteq E(s)$. Then B is valid if and only if there exists an assignment $\mathbf{a}$ for s such that $\{(s, s') \mid s' \in \mathsf{support}(\mathbf{a})\} = B$.*

A consequence of Lemma 1 is that, because we consider only well-formed IMCs, there exists at least one valid subset of outgoing edges from each state.

For each state $s \in S$, we let $Valid(s) = \{B \subseteq E(s) \mid B \text{ is valid}\}$. Note that, in the worst case (when all edges in $E(s)$ are associated with intervals $[0, 1]$), $|Valid(s)| = 2^{|E(s)|} - 1$. Let $Valid = \bigcup_{s \in S} Valid(s)$ be the set of valid sets of the IMC. Given a valid set $B \in Valid$, we let $ValidAssign(B)$ be the set of assignments $\mathbf{a}$ that witness Lemma 1, i.e., all assignments $\mathbf{a}$ such that $\{(s, s') \mid s' \in \mathsf{support}(\mathbf{a})\} = B$. A *witness assignment function* $w : Valid \rightarrow \mathsf{Dist}(S)$ assigns to each valid set $B \in Valid$ an assignment from $ValidAssign(B)$.

Example 1. For the state s_1 of the IMC $\mathcal{O}_2$ of Fig. 2, the valid edge sets are $B_1 = \{(s_1, s_0), (s_1, s_1), (s_1, s_2)\}$ and $B_2 = \{(s_1, s_0), (s_1, s_2)\}$, reflecting the intuition that the edge (s_1, s_1) can be assigned (exactly) probability 0. Note that reducing the right endpoint of (s_1, s_0) to 0.7 would result in B_1 being the only valid set associated with s_1, because B_2 would not be large. An example of a witness assignment function w for state s_1 of $\mathcal{O}_2$ is $w(B_1)(s_0) = 0.7$, $w(B_1)(s_1) = 0.12$ and $w(B_1)(s_2) = 0.18$, and $w(B_2)(s_0) = 0.8$ and $w(B_2)(s_2) = 0.2$.

Qualitative MDP Abstractions. The *qualitative MDP abstraction of $\mathcal{O}$ with respect to witness assignment function w* is the MDP $[\mathcal{O}]_w = (S, \Delta_w)$, where Δ_w is defined by $\Delta_w(s) = \{w(B) \mid B \in Valid(s)\}$ for each state $s \in S$.

3   Qualitative Reachability: UMC Semantics

Qualitative reachability problems can be classified into four categories, depending on whether the probability of reaching the target set T is 0 or 1 for some or for all ways of assigning probabilities to intervals. For the UMC semantics, we consider the computation of the following sets:

- $S_\forall^{0,U} = \{s \in S \mid \forall \mathcal{D} \in [\mathcal{O}]_U \cdot \Pr_s^{\mathcal{D}}(\mathsf{Reach}(T)) = 0\}$;
- $S_\exists^{0,U} = \{s \in S \mid \exists \mathcal{D} \in [\mathcal{O}]_U \cdot \Pr_s^{\mathcal{D}}(\mathsf{Reach}(T)) = 0\}$;
- $S_\exists^{1,U} = \{s \in S \mid \exists \mathcal{D} \in [\mathcal{O}]_U \cdot \Pr_s^{\mathcal{D}}(\mathsf{Reach}(T)) = 1\}$;
- $S_\forall^{1,U} = \{s \in S \mid \forall \mathcal{D} \in [\mathcal{O}]_U \cdot \Pr_s^{\mathcal{D}}(\mathsf{Reach}(T)) = 1\}$.

The remainder of this section is dedicated to showing the following result.

Theorem 1. *The sets $S_\forall^{0,U}$, $S_\exists^{0,U}$, $S_\exists^{1,U}$ and $S_\forall^{1,U}$ can be computed in polynomial time in the size of the IMC.*

Computation of $S_\forall^{0,U}$. The case for $S_\forall^{0,U}$ is straightforward. We compute the state set $S \backslash S_\forall^{0,U} = \{s \in S \mid \exists \mathcal{D} \in [\mathcal{O}]_U \cdot \Pr_s^{\mathcal{D}}(\mathsf{Reach}(T)) > 0\}$, which reduces to reachability on the graph of the IMC according to the following lemma.

Lemma 2. *Let $s \in S$. There exists $\mathcal{D} \in [\mathcal{O}]_U$ such that $\Pr_s^{\mathcal{D}}(\mathsf{Reach}(T)) > 0$ if and only if there exists a path $r \in Paths_*^{\mathcal{O}}(s)$ such that $last(r) \in T$.*

Hence the set $S_\forall^{0,U}$ is equal to the complement of the set of states from which there exists a path reaching T in the graph of the IMC (that is, the graph (S, E)). Given that the latter set of states can be computed in polynomial time, we conclude that $S_\forall^{0,U}$ can be computed in polynomial time.

Computation of $S_\exists^{0,U}$. We show that $S_\exists^{0,U}$ can be obtained by computing the set of states from which there exists a scheduler for which T is reached with probability 0 in the qualitative MDP abstraction $[\mathcal{O}]_w = (S, \Delta_w)$ of $\mathcal{O}$ with respect to some (arbitrary) witness assignment function w.

First we establish that the set of states of $[\mathcal{O}]_w$ for which there exists a scheduler such that T is reached with probability 0 (respectively, probability 1) is equal to the set of states of $\mathcal{O}$ for which there exists a DTMC in $[\mathcal{O}]_U$ such that T is reached with probability 0 (respectively, probability 1).

Lemma 3. *Let* $s \in S$, $\bowtie \in \{<, =, >\}$ *and* $\lambda \in \{0, 1\}$. *There exists* $\mathcal{D} \in [\mathcal{O}]_U$ *such that* $\Pr_s^{\mathcal{D}}(\mathsf{Reach}(T)) \bowtie \lambda$ *if and only if there exists a scheduler* $\sigma \in \Sigma^{[\mathcal{O}]_w}$ *such that* $\Pr_s^{\sigma}(\mathsf{Reach}(T)) \bowtie \lambda$.

In particular, Lemma 3 allows us to reduce the problem of computing $S_{\exists}^{0,U}$ to that of computing the set $\{s \in S \mid \exists \sigma \in \Sigma^{[\mathcal{O}]_w} . \Pr_s^{\sigma}(\mathsf{Reach}(T)) = 0\}$ on $[\mathcal{O}]_w$. As in the case of standard finite MDP techniques (see [11]), we proceed by computing the *complement* of this set, i.e., we compute the set $\{s \in S \mid \forall \sigma \in \Sigma^{[\mathcal{O}]_w} . \Pr_s^{\sigma}(\mathsf{Reach}(T)) > 0\}$. For a set $X \subseteq S$, let $\mathsf{CPre}(X) = \{s \in S \mid \exists \mu \in \Delta_w(s) . \mathsf{support}(\mu) \subseteq X\}$ be the set of states for which there exists a distribution such that all states assigned positive probability by the distribution are in X. Furthermore, we let $\overline{\mathsf{CPre}}(X) = \{s \in S \mid \forall \mu \in \Delta_w(s) . \mathsf{support}(\mu) \cap X \neq \emptyset\}$ be the dual of the CPre operator (i.e., $\overline{\mathsf{CPre}}(X) = S \backslash \mathsf{CPre}(S \backslash X)$), that is the set of states from which it is inevitable to make a transition to X with positive probability. The standard algorithm for computing the set of states of a finite MDP for which all schedulers are such that a set T of target states is reached with probability strictly greater than 0 operates in the following way: starting from $X_0 = T$, we let $X_{i+1} = X_i \cup \overline{\mathsf{CPre}}(X_i)$ for progressively larger values of $i \geq 0$, until we reach a fixpoint (that is, until we obtain $X_{i^*+1} = X_{i^*}$ for some i^*). However, a direct application of this algorithm to $[\mathcal{O}]_w$ would result in an exponential-time algorithm, given that the size of the transition function Δ_w of $[\mathcal{O}]_w$ may be exponential in the size of $\mathcal{O}$. For this reason, we propose an algorithm that operates directly on the IMC $\mathcal{O}$, without needing the explicit construction of $[\mathcal{O}]_w$. We proceed by establishing that CPre can be implemented in polynomial time in the size of $\mathcal{O}$.

Lemma 4. *Let* $s \in S$ *and* $X \subseteq S$. *Then* $s \in \mathsf{CPre}(X)$ *if and only if* *(1)* $E(s, S \backslash X) \subseteq E^{[0,\cdot)}$, *and (2)* $E(s, X)$ *is large. The set* $\mathsf{CPre}(X)$ *can be computed in polynomial time in the size of the IMC* $\mathcal{O}$.

The intuition underlying Lemma 4 is that conditions (1) and (2) encode realisibility and largeness, i.e., validity, of edge set $E(s, X)$. From Lemma 1, their satisfaction means that there exists a distribution in $\Delta_w(s)$ with support set equal to the set of target states of edges in $E(s, X)$. We consider the largest edge set with target states in X, i.e., $E(s, X)$, because taking smaller edge sets with targets in X would make the conditions (1) and (2) more difficult to satisfy.

The final part of Lemma 4 follows from the fact that conditions (1) and (2) in Lemma 4 can be checked in polynomial time in the size of $\mathcal{O}$. Hence our algorithm avoids the construction of the qualitative MDP abstraction $[\mathcal{O}]_w$, and instead consists of direct computation of the sets $X_0 = T$ and $X_{i+1} = X_i \cup S \backslash \mathsf{CPre}(S \backslash X_i)$ for increasing indices i until a fixpoint is reached. Given that a fixpoint must be reached within $|S|$ steps, and the computation of $\mathsf{CPre}(X_i)$ can be done in polynomial time in the size of $\mathcal{O}$, we have that the set $\{s \in S \mid \forall \sigma \in \Sigma^{[\mathcal{O}]_w} . \Pr_s^{\sigma}(\mathsf{Reach}(T)) > 0\}$ can be computed in polynomial time in the size of $\mathcal{O}$. The complement of this set is equal to $S_{\exists}^{0,U}$, as established by Lemma 3, and hence we can compute $S_{\exists}^{0,U}$ in polynomial time in the size of $\mathcal{O}$.

Computation of $S_\exists^{1,U}$. We proceed in a manner analogous to that for the case of $S_\exists^{0,U}$. First we note that, by Lemma 3, we have that $S_\exists^{1,U}$ is equal to the set of states of $[\mathcal{O}]_w$ such that there exists a scheduler for which T is reached with probability 1. Hence, our aim is to compute the set $\{s \in S \mid \exists \sigma \in \Sigma^{[\mathcal{O}]_w} \cdot \Pr_s^\sigma(\mathsf{Reach}(T)) = 1\}$ on $[\mathcal{O}]_w$. We recall the standard algorithm for the computation of this set on finite MDPs [9,10]. Given state sets $X, Y \subseteq S$, we let

$$\mathsf{APre}(Y, X) = \{s \in S \mid \exists \mu \in \Delta_w(s) \cdot \mathsf{support}(\mu) \subseteq Y \land \mathsf{support}(\mu) \cap X \neq \emptyset\}$$

be the set of states for which there exists a distribution such that all states assigned positive probability by the distribution are in Y and there exists a state assigned positive probability by the distribution that is in X. The standard algorithm proceeds by setting $Y_0 = S$ and $X_0^0 = T$. Then the sequence $X_0^0, X_1^0, \cdots$ is computed by letting $X_{i_0+1}^0 = X_{i_0}^0 \cup \mathsf{APre}(Y_0, X_{i_0}^0)$ for progressively larger indices $i_0 \geq 0$ until a fixpoint is obtained, that is, until we obtain $X_{i_0^*+1}^0 = X_{i_0^*}^0$ for some i_0^*. Next we let $Y_1 = X_{i_0^*}^0$, $X_0^1 = T$ and compute $X_{i_1+1}^1 = X_{i_1}^1 \cup \mathsf{APre}(Y_1, X_{i_1}^1)$ for larger $i_1 \geq 0$ until a fixpoint $X_{i_1^*}^1$ is obtained. Then we let $Y_2 = X_{i_1^*}^1$ and $X_0^2 = T$, and repeat the process. We terminate the algorithm when a fixpoint is reached in the sequence $Y_0, Y_1, \cdots$.[1] The algorithm requires at most $|S|^2$ calls to APre. In an analogous manner to CPre in the case of $S_\exists^{0,U}$, we show that APre can characterised by efficiently checkable conditions on $\mathcal{O}$.

Lemma 5. *Let $s \in S$ and let $X, Y \subseteq S$. Then $s \in \mathsf{APre}(Y, X)$ if and only if (1) $E(s, X \cap Y) \neq \emptyset$, (2) $E(s, S \backslash Y) \subseteq E^{[0,\cdot)}$, and (3) $E(s, Y)$ is large. The set $\mathsf{APre}(Y, X)$ can be computed in polynomial time in the size of the IMC $\mathcal{O}$.*

The intuition underlying Lemma 5 is similar to that of Lemma 4.

Hence we obtain an overall polynomial-time algorithm for computing $\{s \in S \mid \exists \sigma \in \Sigma^{[\mathcal{O}]_w} \cdot \Pr_s^\sigma(\mathsf{Reach}(T)) = 1\}$ which, from Lemma 3, equals $S_\exists^{1,U}$.

$S_\forall^{1,U}$. We recall the standard algorithm for determining the set of states for which all schedulers reach a target set with probability 1 on a finite MDP (see [11]): from the set of states of the MDP, we first remove states from which the target state can be reached with probability 0 (for some scheduler), then successively remove states for which it is possible to reach a previously removed state with positive probability. For each of the remaining states, there exists a scheduler that can reach the target set with probability 1.

We propose an algorithm for IMCs that is inspired by this standard algorithm for finite MDPs. Our aim is to compute the complement of $S_\forall^{1,U}$, i.e., the state set $S \backslash S_\forall^{1,U} = \{s \in S \mid \exists \mathcal{D} \in [\mathcal{O}]_U \cdot \Pr_s^\mathcal{D}(\mathsf{Reach}(T)) < 1\}$.

Lemma 6. *Let $s \in S$. There exists $\mathcal{D} \in [\mathcal{O}]_U$ such that $\Pr_s^\mathcal{D}(\mathsf{Reach}(T)) < 1$ if and only if there exists a path $r \in \mathsf{Paths}_*^\mathcal{O}(s)$ such that $\mathsf{last}(r) \in S_\exists^{0,U}$.*

---

[1] Readers familiar with μ-calculus will observe that the algorithm can be expressed using the term $\nu Y \cdot \mu X(T \cup \mathsf{APre}(Y, X))$ [10].

Hence the set $S_\forall^{1,U}$ can be computed by taking the complement of the set of states for which there exists a path to $S_\exists^{0,U}$ in the graph of $\mathcal{O}$. Given that $S_\exists^{0,U}$, and the set of states reaching $S_\exists^{0,U}$, can be computed in polynomial time, we have obtained a polynomial-time algorithm for computing $S_\forall^{1,U}$. Together with the cases for $S_\forall^{0,U}$, $S_\exists^{0,U}$ and $S_\exists^{1,U}$, this establishes Theorem 1.

4 Qualitative Reachability: IMDP Semantics

We now focus on the IMDP semantics, and consider the computation of the following sets:

- $S_\forall^{0,I} = \{s \in S \mid \forall \sigma \in \Sigma^{[\mathcal{O}]_I} . \Pr_s^\sigma(\mathsf{Reach}(T)) = 0\}$;
- $S_\exists^{0,I} = \{s \in S \mid \exists \sigma \in \Sigma^{[\mathcal{O}]_I} . \Pr_s^\sigma(\mathsf{Reach}(T)) = 0\}$;
- $S_\exists^{1,I} = \{s \in S \mid \exists \sigma \in \Sigma^{[\mathcal{O}]_I} . \Pr_s^\sigma(\mathsf{Reach}(T)) = 1\}$;
- $S_\forall^{1,I} = \{s \in S \mid \forall \sigma \in \Sigma^{[\mathcal{O}]_I} . \Pr_s^\sigma(\mathsf{Reach}(T)) = 1\}$.

This section will be dedicated to showing the following result. We note that the cases for $S_\forall^{0,I}$, $S_\exists^{0,I}$ and $S_\exists^{1,I}$ proceed in a manner similar to the UMC case (using either graph reachability or reasoning based on the qualitative MDP abstraction); instead the case for $S_\forall^{1,I}$ requires substantially different techniques.

Theorem 2. *The sets $S_\forall^{0,I}$, $S_\exists^{0,I}$, $S_\exists^{1,I}$ and $S_\forall^{1,I}$ can be computed in polynomial time in the size of the IMC.*

Computation of $S_\forall^{0,I}$. As in the case of UMCs, the computation of $S_\forall^{0,I}$ reduces to straightforward reachability analysis on the graph of the IMC $\mathcal{O}$. The correctness of the reduction is based on the following lemma.

Lemma 7. *Let $s \in S$. There exists $\sigma \in \Sigma^{[\mathcal{O}]_I}$ such that $\Pr_s^\sigma(\mathsf{Reach}(T)) > 0$ if and only if there exists a path $r \in \mathit{Paths}_*^\mathcal{O}(s)$ such that $last(r) \in T$.*

Therefore, to obtain $S_\forall^{0,I}$, we proceed by computing the state set $S \backslash S_\forall^{0,I} = \{s \in S \mid \exists \sigma \in \Sigma^{[\mathcal{O}]_I} . \Pr_s^\sigma(\mathsf{Reach}(T)) > 0\}$, which reduces to reachability on the graph of the IMC according to Lemma 7, and then taking the complement.

Computation of $S_\exists^{0,I}$ and $S_\exists^{1,I}$. In the following we fix an arbitrary witness assignment function w of $\mathcal{O}$. Lemma 8 establishes that $S_\exists^{0,I}$ (respectively, $S_\exists^{1,I}$) equals the set of states of the qualitative MDP abstraction $[\mathcal{O}]_w$ with respect to w for which there exists some scheduler such that T is reached with probability 0 (respectively, probability 1).

Lemma 8. *Let $s \in S$ and $\lambda \in \{0, 1\}$. There exists $\sigma \in \Sigma^{[\mathcal{O}]_I}$ such that $\Pr_s^\sigma(\mathsf{Reach}(T)) = \lambda$ if and only if there exists a scheduler $\sigma' \in \Sigma^{[\mathcal{O}]_w}$ such that $\Pr_s^{\sigma'}(\mathsf{Reach}(T)) = \lambda$.*

Given that we have shown in Sect. 3 that the set of states of the qualitative MDP abstraction $[\mathcal{O}]_w$ for which there exists some scheduler such that T is reached with probability 0 (respectively, probability 1) can be computed in polynomial time in the size of $\mathcal{O}$, we obtain polynomial-time algorithms for computing $S_\exists^{0,I}$ (respectively, $S_\exists^{1,I}$).

Computation of $S_\forall^{1,I}$. This case is notably different from the other three cases for the IMDP semantics, because schedulers that are *not* memoryless may influence whether a state is included in $S_\forall^{1,I}$. In particular, we recall the example of the IMC of Fig. 1: as explained in Sect. 1, we have $s_0 \notin S_\forall^{1,I}$. In contrast, we have $s_0 \in S_\forall^{1,U}$, and s_0 would be in $S_\forall^{1,I}$ if we restricted the IMDP semantics to memoryless (actually finite-memory, in this case) schedulers. For this reason, a qualitative MDP abstraction is not useful for computing $S_\forall^{1,I}$, because it is based on the use of witness assignment functions that assign *constant* probabilities to sets of edges available from states: on repeated visits to a state, the (finite) set of available distributions remains the same in a qualitative MDP abstraction. Therefore we require alternative analysis methods that are not based on the qualitative MDP abstraction. Our approach is based on the notion of end components, which is a standard concept in the field of MDP verification [9]. In this section we introduce an alternative notion of end components, defined solely in terms of states of the IMC, which characterises situations in which the IMC can confine its behaviour to certain state sets with positive probability in the IMDP semantics (for example, the IMC of Fig. 1 can confine itself to state s_0 with positive probability in the IMDP semantics).

An *IMC-level end component* (ILEC) is a set $C \subseteq S$ of states that is strongly connected and such that the total probability assigned to edges that have a source state in C but a target state outside of C can be made to be arbitrarily small (note that such edges must have an interval with a left endpoint of 0). Formally, $C \subseteq S$ is an ILEC if, for each state $s \in C$, we have (1) $E^{\langle +, \cdot \rangle}(s, S \backslash C) = \emptyset$, (2) $\sum_{e \in E(s,C)} \mathsf{right}(\delta(e)) \geq 1$, and (3) the graph $(C, E \cap (C \times C))$ is strongly connected.

Example 2. In the IMC $\mathcal{O}_1$ of Fig. 1, the set $\{s_0\}$ is an ILEC: for condition (1), the edge (s_0, s_1) (the only edge in $E(s_0, S \backslash \{s_0\})$) is not in $E^{\langle +, \cdot \rangle}$, and, for condition (2), we have $\mathsf{right}(\delta(s_0, s_1)) = 1$. In the IMC $\mathcal{O}_2$ of Fig. 2, the set $\{s_0, s_1\}$ is an ILEC: for condition (1), the only edge leaving $\{s_0, s_1\}$ has 0 as its left endpoint, i.e., $\delta(s_1, s_2) = (0, 0.2]$, hence $E^{\langle +, \cdot \rangle}(s_0, \{s_2\}) = E^{\langle +, \cdot \rangle}(s_1, \{s_2\}) = \emptyset$; for condition (2), we have $\mathsf{right}(\delta(s_0, s_0)) + \mathsf{right}(\delta(s_0, s_1)) = 1.6 \geq 1$ and $\mathsf{right}(\delta(s_1, s_0)) + \mathsf{right}(\delta(s_1, s_1)) = 1.3 \geq 1$. In both cases, the identified sets clearly induce strongly connected subgraphs, thus satisfying condition (3).

Remark 1. Both conditions (1) and (2) are necessary to ensure that the probability of leaving C in one step can be made arbitrarily small. Consider an IMC with state $s \in C$ such that $E(s, C) = \{e_1\}$ and $E(s, S \backslash C) = \{e_2, e_3\}$, where $\delta(e_1) = [0.6, 0.8]$, $\delta(e_2) = [0, 0.2]$ and $\delta(e_3) = [0, 0.2]$. Then condition (1) holds but condition (2) does not: indeed, at least total probability 0.2 must be assigned to the edges (e_2 and e_3) that leave C. Now consider an IMC with state $s \in C$ such that $E(s, C) = \{e_1, e_2\}$ and $E(s, S\backslash C) = \{e_3\}$, where $\delta(e_1) = [0, 0.5]$, $\delta(e_2) = [0, 0.5]$ and $\delta(e_3) = [0.1, 0.5]$. Then condition (2) holds (because the sum of the right endpoints of the intervals associated with e_1 and e_2 is equal to 1), but condition (1) does not (because the interval associated with e_3 specifies that probability at least 0.1 must be assigned to leaving C). Note also that if $E(s, C) \subseteq E^{[\cdot, \cdot)} \cup E^{(\cdot, \cdot)}$ (all edges in $E(s, C)$ have right-open intervals) and $\sum_{e \in E(s,C)} \mathrm{right}(\delta(e)) = 1$, there must exist a least one edge in $E(s, S\backslash C)$ by well formedness.

Let $\mathfrak{I}$ be the set of ILECs of $\mathcal{O}$. We say that an ILEC $C \in \mathfrak{I}$ is *maximal* if there does not exist any $C' \in \mathfrak{I}$ such that $C \subset C'$. For a path $\rho \in Paths^{[\mathcal{O}]_\mathrm{I}}(s)$, let $infst(\rho) \subseteq S$ be the states that appear infinitely often along ρ, i.e., for $\rho = s_0 \mu_0 s_1 \mu_1 \cdots$, we have $infst(\rho) = \{s \in S \mid \forall i \in \mathbb{N} . \exists j > i . s_j = s\}$. We present a result for ILECs that is analogous to the fundamental theorem of end components of [9]: the result specifies that, with probability 1, a scheduler of the IMDP semantics of $\mathcal{O}$ must confine itself to an ILEC.

Lemma 9. *For $s \in S$ and $\sigma \in \Sigma^{[\mathcal{O}]_\mathrm{I}}$, we have $\mathrm{Pr}^\sigma_s(\{\rho \mid infst(\rho) \in \mathfrak{I}\}) = 1$.*

We now show that there exists a scheduler that, from a state within an ILEC, can confine the IMC to the ILEC with positive probability. This result is the ILEC analogue of a standard result for end components of finite MDPs that specifies that there exists a scheduler that, from a state of an end component, can confine the MDP to the end component with probability 1 (see [2,9]). In the case of IMCs and ILECs, it is not possible to obtain an analogous result for probability 1; in the example of Fig. 1, the singleton set $\{s_0\}$ is an ILEC, but it is not possible to find a scheduler that remains in s_0 with probability 1, because with each transition the IMC goes to s_1 with positive probability. For our purposes, it is sufficient to have a result stating that, from an ILEC, the IMC can be confined to the ILEC with positive probability.

Lemma 10. *Let $C \in \mathfrak{I}$ and $s \in C$. There exists $\sigma \in \Sigma^{[\mathcal{O}]_\mathrm{I}}$ such that $\mathrm{Pr}^\sigma_s(\{\rho \mid \rho \notin \mathsf{Reach}(S\backslash C) \wedge infst(\rho) = C\}) > 0$.*

The key point of the proof of Lemma 10 is the definition of a scheduler that assigns progressively decreasing probability to all edges in $E^{(0, \cdot)}$ that leave ILEC C, in such a way as to guarantee that the IMC is confined in C with positive probability. This is possible because condition (2) of the definition of ILECs specifies that there is no lower bound on the probability that must be assigned to edges that leave C. Furthermore, the scheduler is defined so that the remaining probability at each step that is assigned between all edges that stay in C is always no lower than some fixed lower bound; this characteristic of the scheduler, combined with the fact that we remain in C with positive probability and the fact that C is strongly connected, means that we visit all states of C with positive probability under the defined scheduler.

Let $U_{\neg T} = \bigcup\{C \in \mathfrak{I} \mid C \cap T = \emptyset\}$ be the union of states of ILECs that do not contain states in T. Using Lemmas 9 and 10 in a standard way, we can show that the existence of a scheduler of $[\mathcal{O}]_I$ that reaches T with probability strictly less than 1 is equivalent to the existence of a path in the graph of $\mathcal{O}$ that reaches $U_{\neg T}$.

Proposition 1. *Let $s \in S$. There exists $\sigma \in \Sigma^{[\mathcal{O}]_I}$ such that $\mathrm{Pr}_s^\sigma(\mathsf{Reach}(T)) < 1$ if and only if there exists a finite path $r \in \mathit{Paths}_*^{\mathcal{O}}(s)$ such that $\mathit{last}(r) \in U_{\neg T}$.*

Hence we identify the set $S_{\forall}^{1,I}$ by computing the complement of $S_{\forall}^{1,I}$, i.e., the set $S \backslash S_{\forall}^{1,I} = \{s \in S \mid \forall \sigma \in \Sigma^{[\mathcal{O}]_I} . \mathrm{Pr}_s^\sigma(\mathsf{Reach}(T)) < 1\}$. Using Proposition 1, this set can be computed by considering reachability on the graph of $\mathcal{O}$ of the set $U_{\neg T}$. The set $U_{\neg T}$ can be computed in polynomial time in the size of $\mathcal{O}$ in a manner similar to the computation of maximal end components of MDPs (see [2,9]). First we compute all strongly connected components $(C_1, E \cap (C_1 \times C_1)), \cdots, (C_m, E \cap (C_m \times C_m))$ of the graph $(S \backslash T, E \cap ((S \backslash T) \times (S \backslash T)))$ of $\mathcal{O}$. Then, for each $1 \leq i \leq m$, we remove from C_i all states for which conditions (1) or (2) in the definition of ILECs do *not* hold with respect to C_i (these conditions can be checked in polynomial time for each state), to obtain the state set C_i'. Next, we compute the strongly connected components of the graph $(C_i', E \cap (C_i' \times C_i'))$, and for each of these, repeat the procedure described above. We terminate the algorithm when it is not possible to remove a state (via a faliure to satisfy a least one of the conditions (1) and (2) in the definition of ILECs) from any generated strongly connected component. The generated state sets of the strongly connected components obtained will be be the maximal ILECs that do not contain any state in T, and their union is $U_{\neg T}$. Hence the overall algorithm for computing $S_{\forall}^{1,I}$ is in polynomial time in the size of $\mathcal{O}$.

5   Conclusion

We have presented algorithms for qualitative reachability properties for open IMCs. In the context of qualitative properties of system models with *fixed* probabilities on their transitions, probability can be regarded as imposing a fairness constraint, i.e., paths for which a state is visited infinitely often and one of its successors is visited only finitely often have probability 0. In open IMCs, the possibility to make the probability of a transition converge to 0 in the IMDP semantics captures a different phenomenon, which is key for problems concerning the minimum reachability probability being compared to 1. We conjecture that finite-memory strategies are no more powerful than memoryless strategies for this class of problem. For the three other classes of qualitative reachability problems, we have shown that the UMC and IMDP semantics coincide. We note that the algorithms presented in this paper require some numerical computation (a sum and a comparison of the result with 1 in the CPre, APre and ILEC computations), but these operations are simpler than the polynomial-time solutions for quantitative properties of (closed) IMCs in [6,17]. Similarly, the CPre and APre operators are simpler than the polynomial-time step of value iteration used in the context of quantitative verification of [12]. For the IMDP semantics, our methods give directly a P-complete algorithm for the qualitative fragment of the temporal logic PCTL [13]. Future work could consider quantitative properties and ω-regular properties, and applying the results to develop qualitative reachability methods for interval Markov decision processes or for higher-level formalisms such as clock-dependent probabilistic timed automata [19].

References

1. Baier, C., de Alfaro, L., Forejt, V., Kwiatkowska, M.: Model checking probabilistic systems. Handbook of Model Checking, pp. 963–999. Springer, Cham (2018). https://doi.org/10.1007/978-3-319-10575-8_28
2. Baier, C., Katoen, J.-P.: Principles of Model Checking. MIT Press, Cambridge (2008)
3. Caillaud, B., Delahaye, B., Larsen, K.G., Legay, A., Pedersen, M.L., Wasowski, A.: Constraint Markov chains. Theor. Comput. Sci. **412**(34), 4373–4404 (2011)
4. Chakraborty, S., Katoen, J.-P.: Model checking of open interval Markov chains. In: Gribaudo, M., Manini, D., Remke, A. (eds.) ASMTA 2015. LNCS, vol. 9081, pp. 30–42. Springer, Cham (2015). https://doi.org/10.1007/978-3-319-18579-8_3
5. Chatterjee, K., Sen, K., Henzinger, T.A.: Model-checking ω-regular properties of interval Markov chains. In: Amadio, R. (ed.) FoSSaCS 2008. LNCS, vol. 4962, pp. 302–317. Springer, Heidelberg (2008). https://doi.org/10.1007/978-3-540-78499-9_22
6. Chen, T., Han, T., Kwiatkowska, M.: On the complexity of model checking interval-valued discrete time Markov chains. Inf. Process. Lett. **113**(7), 210–216 (2013)
7. Courcoubetis, C., Yannakakis, M.: The complexity of probabilistic verification. J. ACM **42**(4), 857–907 (1995)
8. Daws, C.: Symbolic and parametric model checking of discrete-time Markov chains. In: Liu, Z., Araki, K. (eds.) ICTAC 2004. LNCS, vol. 3407, pp. 280–294. Springer, Heidelberg (2005). https://doi.org/10.1007/978-3-540-31862-0_21
9. de Alfaro, L.: Formal verification of probabilistic systems. Ph.D. thesis, Stanford University, Department of Computer Science (1997)
10. Alfaro, L.: Computing minimum and maximum reachability times in probabilistic systems. In: Baeten, J.C.M., Mauw, S. (eds.) CONCUR 1999. LNCS, vol. 1664, pp. 66–81. Springer, Heidelberg (1999). https://doi.org/10.1007/3-540-48320-9_7
11. Forejt, V., Kwiatkowska, M., Norman, G., Parker, D.: Automated verification techniques for probabilistic systems. In: Bernardo, M., Issarny, V. (eds.) SFM 2011. LNCS, vol. 6659, pp. 53–113. Springer, Heidelberg (2011). https://doi.org/10.1007/978-3-642-21455-4_3
12. Haddad, S., Monmege, B.: Interval iteration algorithm for MDPs and IMDPs. Theor. Comput. Sci. **735**, 111–131 (2018)
13. Hansson, H., Jonsson, B.: A logic for reasoning about time and reliability. Formal Aspects Comput. **6**(5), 512–535 (1994)
14. Jonsson, B., Larsen, K.G.: Specification and refinement of probabilistic processes. In: 1991 Proceedings of LICS, pp. 266–277. IEEE Computer Society (1991)
15. Kozine, I.O., Utkin, L.V.: Interval-valued finite Markov chains. Reliable Comput. **8**(2), 97–113 (2002)
16. Lanotte, R., Maggiolo-Schettini, A., Troina, A.: Parametric probabilistic transition systems for system design and analysis. Formal Aspects Comput. **19**(1), 93–109 (2007)

17. Puggelli, A., Li, W., Sangiovanni-Vincentelli, A.L., Seshia, S.A.: Polynomial-time verification of PCTL properties of MDPs with convex uncertainties. In: Sharygina, N., Veith, H. (eds.) CAV 2013. LNCS, vol. 8044, pp. 527–542. Springer, Heidelberg (2013). https://doi.org/10.1007/978-3-642-39799-8_35
18. Sen, K., Viswanathan, M., Agha, G.: Model-checking Markov chains in the presence of uncertainties. In: Hermanns, H., Palsberg, J. (eds.) TACAS 2006. LNCS, vol. 3920, pp. 394–410. Springer, Heidelberg (2006). https://doi.org/10.1007/11691372_26
19. Sproston, J.: Probabilistic timed automata with clock-dependent probabilities. In: Hague, M., Potapov, I. (eds.) RP 2017. LNCS, vol. 10506, pp. 144–159. Springer, Cham (2017). https://doi.org/10.1007/978-3-319-67089-8_11
20. Sproston, J.: Qualitative reachability for open interval Markov chains. CoRR (2018)
21. Vardi, M.: Automatic verification of probabilistic concurrent finite-state programs. In: 1985 Proceedings of FOCS, pp. 327–338. IEEE Computer Society (1985)

Author Index

Alexandre dit Sandretto, Julien 1

Bensalem, Saddek 30
Boneva, Iovka 117

Cook, Matthew 103

Day, Joel D. 15
de Oliveira, Steven 30

Ganesh, Vijay 15

Habermehl, Peter 30
Hague, Matthew 45
He, Paul 15

Jančar, Petr 59

Lisitsa, Alexei 75
Lohrey, Markus 87

Manea, Florin 15

Neary, Turlough 103
Niehren, Joachim 117
Nowotka, Dirk 15

Osička, Petr 59

Penelle, Vincent 45
Prevosto, Virgile 30

Sakho, Momar 117
Sawa, Zdeněk 59
Skrzypczak, Michał 133
Sproston, Jeremy 146

Wan, Jian 1

Printed in the United States
By Bookmasters